W9-BMK-470

DATE DUE

BUSINESS & CONSTRUCTION

FERGUSON

GREEN CAREERS

BUSINESS & CONSTRUCTION

PAMELA FEHL

Ferguson Publishing
An imprint of Infobase Publishing

Green Careers: Business and Construction

Copyright © 2010 by Infobase Publishing

Ferguson
An imprint of Infobase Publishing
132 West 31st Street
New York NY 10001

Library of Congress Cataloging-in-Publication Data
Fehl, Pamela.
    Business and construction / [Pamela Fehl].
      p. cm. — (Green careers)
    Includes bibliographical references and index.
    ISBN-13: 978-0-8160-8149-3 (hbk. : alk. paper)
    ISBN-10: 0-8160-8149-2 (hbk. : alk. paper)
    1. Construction industry—Juvenile literature. 2. Business—Vocational guidance—Juvenile literature. 3. Environmentalism—Juvenile literature. I. Title.
    HD9715.A2F44 2010
    311.702dc22   2009037041

Ferguson books are available at special discounts when purchased in bulk quantities for businesses, associations, institutions, or sales promotions. Please call our Special Sales Department in New York at (212) 967-8800 or (800) 322-8755.

You can find Ferguson on the World Wide Web at http://www.fergpubco.com

Text design and composition by Annie O'Donnell
Cover printed by Bang Printing, Brainerd, MN
Book printed and bound by Bang Printing, Brainerd, MN
Date printed: March 2010
Printed in the United States of America

10 9 8 7 6 5 4 3 2 1

Contents

Introduction

Choosing a career path is challenging. On top of everything else that you have to deal with in high school, you face the big task of figuring out which colleges to apply to and what subject or subjects to major in. Down the road, you hope all of your hard work leads you to a job you love, because the alternative is, well, it's just not very inviting. That you have this book in your hand is an excellent sign. You're off to a great start because green-collar jobs have a bright future. Granted, neglect and abuse of the environment is what's inspiring the burgeoning green business industry. But on the positive side, more opportunities exist than ever before for you to bring your knowledge, talent, and passion to the workplace, and to help improve things. If you need more good news to reassure you that you're looking in the right direction, the U.S. Department of Labor forecasts good employment growth for many of the jobs in the green-collar industry.

The focus on global warming, concerns about practicing and operating business in ways that sync more closely with nature and enhance, rather than destroy, communities, and the proliferation of recycling and litter- and pollution-reduction programs form the basis of the emerging "green" economy. What's more, federal, state, and local governments, as well as those around the globe, are enacting stricter regulations regarding carbon emissions in an effort to reduce greenhouse gases in the atmosphere. Many companies no longer have a choice in some of the ways they conduct business—laws are forcing them to get their acts together and comply. Sustainable business practices are becoming standardized. Eco-businesses that once stood out as unique in years past are now the norm.

The *Business and Construction* volume of the Green Careers series highlights a small sampling of jobs to give you an idea of the different areas in which you can work in this industry. They are: architects, green construction; corporate climate strategists; eco-investors; environmental bankers; environmental economists; green builders; green products manufacturers; green recruiters; grounds maintenance workers; land acquisition professionals; landscape architects; land trust or preserve managers; managers and owners, green business; salespeople, green product; and surveyors.

Each job profile is broken down into sections that will help give a better grasp of the job and make it easier for you to see if your interests and skills line up with what's required.

- **Quick Facts** is exactly that: a snapshot of basic info, ranging from school subjects to job future.
- The **Overview** is a short summary of the job, with a brief description of responsibilities.
- Some jobs in the now "green" business and construction field have been around for a long time, and others are still fairly new and evolving—but they all originated somewhere; the **History** section tells you how and why they began.
- **The Job** features a deeper exploration of the work, with some profiles including comments and insights from people who have direct experience in the field.
- You can map out which classes to start taking in high school and beyond by reading the **Requirements** section. You'll also find out if your skills and character traits match those needed to enjoy and succeed at the work.
- In **Exploring** you'll find recommendations for ways to learn more about the job on your own, whether it's suggestions for specific books and magazines to read, or Web sites and online videos to check out.
- **Employers** focuses on types of industries and companies that hire the worker that's featured, and may include statistics regarding the number of professionals employed in the United States, and the states and/or cities in which most professionals are concentrated. Statistics are often derived from the U.S. Department of Labor (DoL), the National Association of Colleges and Employers, and professional industry-related associations.
- **Starting Out** gives you tips on the steps you can start taking now toward finding a job in this field.
- Career paths vary for green business and construction professionals. The **Advancement** section sheds light on the different jobs and specialties professionals can move up to.
- **Earnings** gives you salary ranges for the specific job and closely related jobs. Information is based on surveys conducted by the DoL, and sometimes from such employment specialists as Salary.com.
- The **Work Environment** section describes the typical surroundings and conditions of employment. Also discussed are typical hours worked, any seasonal fluctuations, and the stresses and strains of the job.

❧   Does it have a bright future, or is it a risky job? Most jobs depend on the economy. When things are up, jobs abound. When things slow down, fewer jobs exist and competition heats up. The **Outlook** section tells you the forecast for the job. It may be based on results from DoL surveys, or findings from professional associations' industry studies.

❧   Each profile ends with **For More Information**, providing you with listings and contact information for professional associations you may want to join and other resources you can use to learn more about the job.

Questions and ideas may crop up as you read through each profile—be sure to keep a notebook or your laptop handy so you can take notes. Also bear in mind that the information in this book is based on industry research and interviews with experts conducted at the time of publication. Job requirements, salary ranges, industry outlooks, and even Web addresses change regularly. Use the resources mentioned in the profiles to find the most current information about the jobs that interest you most.

Job exploration gives you the opportunity to ask lots of questions, meet new people, and learn at least one thing you didn't know before. Now's your chance to dig deeper until you find something that clicks. One final tip: Enjoy the process, and have fun!

Architects, Green Construction

OVERVIEW

Green construction architects (sometimes called *eco-architects*) plan and design buildings that use energy, water, and materials efficiently, with as little impact as possible on the environment and on human health. They create green buildings, which operate in harmony with their natural surroundings. Green construction architects study the environment of the site location, making sure local resources are used in the best ways. For example, they may use recycled metal or stone, lumber from forests that have been certified as sustainably managed, and recycled energy products such as foundry sand and demolition debris. Homeowners, municipalities, corporations, and sometimes even other architects hire green construction architects to help them create green buildings and green building programs that adhere to local building codes, climate, and economy.

HISTORY

For centuries, people have called upon architects to create functional, aesthetically pleasing buildings for homes, businesses, education, religion, sports, and entertainment. And the practice of creating structures that sync with the environment and climate dates to even earlier times.

The ancient Roman aqueducts were early examples of function combined with beauty. Roman architects engineered these waterways to run underground for 260 miles, choosing locations based on gradients and gravity. Thirty miles of the waterways were visible above the ground, consisting of massive arched structures. The waterways—created from stone, brick, and cements made from volcanic ash, and running through metal pipes—carried water to public fountains, baths, and private villas. In ancient times, 11 aqueducts supplied the city of Rome with over 1 million cubic meters of water each day, which would be enough to supply water to 3.5 million people even today.

Ancient adobe houses in the Southwest are another good example of building "green." Back then, the intention in construction had nothing to do with preserving the environment, but came about due to the availability of materials and economy and convenience. Settlers did not have access to wood or masonry materials, but clay, sand, and water were abundant. They combined these materials and poured them into wooden forms, often using their feet to shape bricks. Once the bricks had dried, they took them out, stacked them, and within one month, the bricks became as hard as cement. People favored adobe because it is adaptable to temperature and climate changes, which the Southwest is particularly known for. And while in the past adobe homes connoted poor or low-income living, today people of all income levels live in them. Adobe construction is on the rise because of its practicality and rustic beauty. And while the construction process has been industrialized, many people are still choosing to build their adobe homes with their own hands (and feet).

THE JOB

Architects who specialize in green construction help clients create buildings with various environmentally sound features. They create structures that are healthy and non-toxic, energy efficient, resource and water conserving, and site preservation/restoration-friendly.

Architects tour the University of Texas Health Science Center at Houston's new school of nursing building. The building is one of the largest and most sophisticated green buildings in the Southwest. *AP Photo/David J. Phillip*

Green construction architects may either own their own firms or work for larger firms. The Bureau of Labor Statistics cites that 20 percent of all architects are self-employed, which is more than twice the proportion for all other types of occupations. Clients who hire green architects need residential, commercial, or industrial structures. It may be a new structure, a remodel, an addition to a home, or a reconstruction of an existing building.

"Green architects design buildings that act like sailboats," Todd Jersey Architects, a San Francisco-based firm, explain on their Web site (http://www.toddjerseyarchitec ture.com). "Sailboats are designed to use wind (a form of solar energy) to provide their energy. Green buildings are also designed to use the sun to provide their energy. Conventional architecture firms, on the other hand, design their buildings like powerboats. Both powerboats and conventional buildings are designed to be totally reliant on fossil fuels to provide their energy."

Jeremy Levine, who founded the Los Angeles-based architectural firm Jeremy Levine Design (http://www.jeremylevine.com) in 2002, breaks the job down into seven steps: 1) meet client; 2) sketch concepts; 3) mock up the design; 4) project development; 5) project permit; 6) collect contractors' bids; and 7) construction supervision.

Step 1: Meet the Client

Communication is a crucial element of this job, and the work depends strongly on building relationships. "It's a big responsibility," Levine says. "It's also very exciting. People give you their dreams—they're entrusting you with the most important thing in their life. They tell you their vision, and it's up to you to pull out the specifics to get to the functional requirements, such as what kind of space they want (four-bedroom? two bedroom?). And do they want all recycled materials? Will they want wind energy? You'll have to get suppliers for these types of things." Based on the conversation, and the client's budget, eco-architects then do a lot of research—such as speaking with suppliers and consultants, learning more about the community and location of the building—and bring their findings back to the client for further discussion and negotiation.

Educating the client, as well as themselves, is a big part of the eco-architect's work. "You become a Green Ambassador," Levine says. "With normal architecture, people know what a house looks like. But when you say 'gray water recycling' to them, they say, 'What the heck is *that?*' You educate people about the possibilities." (See the sidebar "Green Building Terms" for the definition of gray water recycling.)

Step 2: Sketch Concepts

In this phase, eco-architects walk around the site sketching and drawing their ideas.

They visit the site multiple times, share the sketches with clients, and rework them according to specifications. They take measurements, photos, and video of the location for use in the next step.

Step 3: Mock up the Design

Using all the information gathered during the previous phase, eco-architects get down to work in their studio and create 3-D models using computer graphics. Doing this gives them a series of ways to make the house green; for example, they can determine how and where to use solar energy.

Step 4: Project Development

By this stage, it's time to bring in the team to carry out the project. Eco-architects bring in either staff or consultants, who may be:

- structural engineers
- mechanical engineers
- solar energy consultants
- wind energy consultants
- gray water recycling consultants
- landscaping architects
- arborists
- surveyors

If it's a new construction, the team works together to decide such things as where the best place is to put the house without cutting down trees. If it's a rebuild, they consider such things as where best to situate windows for solar energy use, or where to put trees to create shade for cooling the house.

Step 5: Project Permit

Eco-architects need building permits to execute their projects. Once the details are finalized and approved by the client, the eco-architect presents the project to the city. City engineers review it and write notes regarding zoning regulations and building codes. Levine says it's "like a game, trying to figure out how to make it happen." And that's where permit expeditors come into play— these are consultants who know the ins and outs of the city building permit process. Architects often hire them to help them speed things up.

Step 6: Collect Contractors' Bids

Here, the eco-architect gathers all of the bids from contractors and brings them to the client. Many architects also partner with other firms on their projects.

Step 7: Construction Supervision

Depending on the client, eco-architects may also make site visits throughout the construction phase. Project managers or construction supervisors make sure materials that have been specified are being used and that work is being done according to approved plans and schedules.

In addition to project work and coordinating teams, eco-architects have to pay attention to everything else related to their work, from marketing and promoting their services to maintaining their offices. They may have someone handling their bills and payroll, or, if they're a one- or two-person office, they may be balancing the accounts themselves. If they have a staff, they may be responsible for hiring employees, supervising personnel, and conducting performance reviews. Their day-to-day responsibilities will depend on the size of their firm.

REQUIREMENTS

High School

Students interested in working as eco-architects need strong analytical, math, artistic, and language skills. "The more well-rounded you are, the better off you are," says eco-architect Marilyn Miller Farmer, LEED AP (Leadership in Energy and Environmental Design, Accredited Professional), of Habitat Studio (http://habitatdesigns.com), based in Los Osos, California. "Architecture bridges the gap between art and science, so explore all: physics, science, math, and art."

A passion for the environment and concern about human health are also required. The job entails working independently to create designs, as well as consulting with clients and collaborating on project teams. While in high school, pursue a rigorous academic program in math, science (including physics), English, art, and foreign languages. If available, take as many advanced or honors class in these subjects as possible.

Postsecondary Training

To practice architecture professionally, most states require that students have architectural degrees from one of the 114 schools with

GREEN BUILDING TERMS

Energy Star   The U.S. Environmental Protection Agency (EPA) started the volunteer-labeling program Energy Star in 1992 to help reduce greenhouse gas emissions by identifying and promoting energy-efficient products. Now a joint program of the EPA and the U.S. Department of Energy, the Energy Star designation has grown to encompass appliances, lighting, and electronics as well as homes and commercial and industrial buildings.

FSC-certified wood   The Forest Stewardship Council (FSC) certifies forests that meet criteria for exhibiting good sustainability and management practices. Wood from FSC-certified forests is quickly renewable, and the forests operate in ways that limit the impact on the environment and people.

gray water recycling   Gray water is all the nontoilet wastewater produced in the average household, such as water from bathtubs, showers, sinks, washing machines, and dishwashers. After the toxins and other dangerous matter are removed, the gray water is then typically used (recycled) to water nonedible gardens. Benefits of recycling gray water include reduced need for fresh water, less energy and fewer chemicals, and better growth for plants because of the small bits of compost usually found in gray water.

green roof   Also known as living roofs and eco-roofs, green roofs are roofs that are covered with living vegetation. A waterproof membrane protects the interior of the building, and soil and plants are placed on top of this membrane. Green roofs absorb rainwater, cool the temperature inside buildings, and provide habitats for wildlife.

net zero energy house   Through the use of solar energy and other energy-saving strategies, this type of house produces as much energy as it uses, often producing more than it needs.

passive climate control   In this type of sustainable building, nature, and natural materials, help cool and heat the interiors of homes and buildings. To make the best use of the environment and climate, eco-architects consider the location of the structure (e.g., its orientation to sun and winds); the available natu-

(continues)

(continued)

ral shading (trees, hills, other buildings); and building materials (for instance, lightweight walls in tropical areas). Solar energy is a good example of passive climate control: solar panels capture energy from the sun to heat water and interior spaces of buildings.

Xeriscaping (or drought-tolerant landscaping) Originated in Colorado, xeriscaping is a water-conservative approach to landscaping for areas that are particularly dry. (*Xeros* is the Greek word for dry.) Native plants that are climate-friendly are typically used (cacti, grasses, and wildflowers, for example) to create appealing landscapes that need less than half the water required for traditional landscapes.

degree programs accredited by the National Architectural Accrediting Board (NAAB), according to the U.S. Department of Labor. Students have the choice of three types of architecture degrees: five-year bachelor's degree (for those without previous architectural training); two-year master's degree (for those with an undergraduate degree in architecture or related subject); or a three- or four-year master's degree for those with a degree in another subject. Course work will include architectural history and theory, building design (with an emphasis on computer-aided design), physical sciences, and liberal arts.

Some of the schools offering degree programs in environmental architecture include California Polytechnic State University in San Luis Obispo, California, (http://www.arch.calpoly.edu); the College of Environmental Design at University of California, Berkeley (http://www.ced.berkeley.edu); Boston Architectural College (http://www.the-bac.edu); and the University of Colorado (http://www.colorado.edu).

Practicing architects usually do not need to have graduate degrees, although those whose work focuses on research, teaching, or certain specialties may be required to have post-professional degrees.

Architecture graduates must complete three years of training to be eligible for the licensing exam. Depending on the state, training can be with a licensed architect or in the office of related professionals, such as engineers or general contractors.

Certification or Licensing

Most states and the District of Columbia require individuals to be licensed to practice architecture professionally. To take the licensing exam (known as the Architect Registration Examination), one must have a degree in architecture from a NAAB-accredited program and three years of training. The exam consists of nine divisions, all of which the individual must pass to be licensed.

Many states also require practicing architects to take continuing education courses to maintain their license. They may do this through universities, associations, conferences, or self-study courses.

Many architects also volunteer to get certified by the National Council of Architectural Boards as a way to further their careers. Certification also eases the licensing process across the states.

Other Requirements

The ability to think creatively and visually is key in this job. According to Levine, solid drawing and photography skills are essential. "You need to be a wizard at computers," he says. "And be very, very good at CAD [computer-aided design] and Photoshop. There's nothing this job doesn't encompass—it requires multiple, overlapping skills."

EXPLORING

The topic "green" is everywhere, so you will find tons of information on the subject in all variety of media. Read magazines such as *Architect, Architectural Record, Art Forum, Dwell, Green Source*, and *Natural Home*. You can also learn more about sustainable design and technologies by subscribing (for free) to the quarterly newsletter *High Performing Buildings* (http://www.hpbmagazine.org/).

Marilyn Miller Farmer summed up her advice for exploring this field: "Travel and sketch." She says, "Beyond school, the best education is to explore other environments, other cultures. It gives you a broader experience."

EMPLOYERS

Eco-architects are either self-employed or work for architectural, engineering, or other related services firms. In 2006 about 132,000 architects were employed in the United States, according to the U.S. Department of Labor. Some work for nonresidential and residential construction firms, and for government agencies involved in housing, community planning, or construction of government buildings.

STARTING OUT

Many eco-architects get their start through internships while they are still in school. Working in an architectural firm that focuses on sustainable design and building is one of the best ways to learn firsthand about the field. Architectural interns may help with such areas as designing part of a project, preparing architectural documents or drawings, building models, and preparing construction drawings (using computer-aided design). They may also research building codes and materials or write specifications for building materials, installation criteria, the quality of finishes, and other related details. Contact the American Institute of Architects (listed at the end of this article) to locate green architectural firms near where you live.

ADVANCEMENT

Eco-architects who are self-employed advance over the years by increasing their knowledge of sustainable design; honing their skills in specific areas; managing more projects; growing their reputation and client base; and expanding their staff and offices. Architects who work for firms advance in similar ways: by being given more responsibilities over the years until they are managing multiple projects, coordinating and overseeing staff and consultants, and eventually becoming partners in established firms.

EARNINGS

In 2008 half of all architects had median incomes of $70,320, according to the U.S. Department of Labor. The lowest 10 percent earned an average of $41,320, and the highest 10 percent earned $119,220 or more. The work is dependent on economy and geography. Some parts of the country do not have many building projects at all; other areas may have only residential building projects. During recessions, office and retail construction typically diminishes, while institutional building construction (e.g., schools, hospitals, nursing homes, correctional facilities) usually remains unaffected.

Architects who work for firms have fringe benefits such as paid vacations and health insurance coverage, as well as reimbursement for school tuition and continuing education fees. Staff and self-employed architects often bolster their incomes by lecturing and consulting, and writing for books, magazines, and Web sites. Some also consult with, and may appear in, television shows (e.g., the Discovery Channel's *Renovation Nation*, and on the Planet Green Network, which featured the construction of Levine's North Eagle Rock house in the summer of 2008).

WORK ENVIRONMENT

Eco-architects mostly work in their offices, but they travel for site visits, client meetings, and, if needed, construction supervision. Outdoor conditions will vary depending on where the building project is located. Architects may need to take measurements in freezing temperatures and snowy conditions; during torrential rains; or in blistering heat. Work hours will also vary depending on the project deadline—for some weeks, and even months, this can mean a 40-hour work week; while at other times eco-architects may need to put in 60 or more hours per week, including evenings and weekends.

OUTLOOK

The U.S. Bureau of Labor Statistics forecasts that employment of architects overall is expected to grow faster than the average for all occupations through 2016. Eco-architects will be in particular demand due to the population's increasing awareness about environmental issues. The economy and the environment will drive the rise in need for eco-architects. People are becoming more educated about ways to build sustainably, and they are learning that by doing so they not only help conserve resources but also save money in the long term. All of these factors will continue to motivate them to hire architects who specialize in green design and construction.

FOR MORE INFORMATION

Since 1857, this professional membership association has been a resource for licensed architects, emerging professionals, and allied architectural partners.

American Institute of Architects
1735 New York Avenue, NW
Washington, DC 20006-5292
Tel: 800-AIA-3837

Email: infocentral@aia.org
http://www.aia.org

Find out more about certified forests and green building by visiting
Forest Stewardship Council–U.S. (FSC-US)
212 Third Avenue North, Suite 280
Minneapolis, MN 55401-1442
Tel: 612-353-4511
Email: info@fscus.org

This organization offers programs and projects centered on high-performance buildings.
Sustainable Buildings Industry Council
1112 16th Street, NW, Suite 240
Washington, DC 20036-4818
Tel: 202-628-7400
Email: sbic@sbicouncil.org
http://www.sbicouncil.org

To learn more about Energy Star products, homes, and buildings, visit the Energy Star Web site.
U.S. Environmental Protection Agency
Energy Star Hotline (6202J)
1200 Pennsylvania Avenue, NW
Washington, DC 20460-0001
Tel: 888-782-7937
http://www.energystar.gov

Composed of more than 19,500 organizations across the building industry, this council aims to expand sustainable building practices. To learn more about membership, courses, and e-newsletters, contact
U.S. Green Building Council
2101 L Street, NW, Suite 500
Washington, DC 20037-1599
Tel: 800-795-1747
Email: info@usgbc.org
http://www.usgbc.org

Corporate Climate Strategists

OVERVIEW

Corporate climate strategists advise companies on how to reduce their "footprint" on the environment and run their businesses in a more sustainable way. Strategists review current company practices and environmental policies; meet with scientists, environmental policy makers, government officials, consultants, corporate executives, and employees; attend workshops and conferences on corporate sustainability; and create corporate environmental strategies to reduce emissions and increase investment in clean technology.

HISTORY

Human beings have "very likely" been the primary cause of changes in the climate since 1950, according to the United Nations' Intergovernmental Panel on Climate Change (IPCC) in its 2007 report, "The Physical Basis of Climate Change." And we have actually been affecting the environment for far longer than this. IPCC cites that human activity has contributed strongly to climate change for almost three centuries, starting at the time of the Industrial Revolution in 1750.

Global warming is the result of the increase of global greenhouse gas (GHG) emissions—most notably carbon dioxide, methane, and nitrous oxide. Because the concentration of these GHGs has grown too large to escape into space, heat gets trapped in the atmosphere, which then causes temperatures to rise. This has a domino effect: Rising temperatures affect the climate and weather patterns, which can alter the length of seasons, which can cause more storms and coastal flooding, all of which can impact species' migration patterns as well as survival. The use of fossil fuels (e.g., coal and oil) has contributed to the rise of carbon dioxide in the atmosphere. Changes in the climate are linked to emissions from automobiles, airplanes, power plants, buildings, agricultural practices, and even deforestation (fewer trees means less carbon dioxide being converted to oxygen).

People started tuning in to the impact of pollution and litter on the environment in the 1960s and early 1970s. Environmental awareness was raised through natural history writer Rachel Carson's book, *Silent Spring* (published in 1962), which pointed out the harmful effects of insecticides. According to the Natural Resources Defense Council, another highlight came in 1963, when New York City utility company Consolidated Edison announced its plan to build a power plant on Storm King Mountain, located near the Hudson River. Longtime residents launched an opposition campaign and organized a conservation group to preserve the scenic environment. This was the first conservation group permitted to sue on the behalf of public interest, and the court's decision in its favor—*Scenic Hudson Preservation Conference v. Federal Power Commission (1965)*—set a legal landmark.

The first Earth Day (April 22, 1970) brought environmental issues into the global spotlight, triggering the passing of thousands of environmental laws throughout the 1970s and 1980s, including the Federal Water Pollution Control Act and the Environmental Pesticide Control Act. And, unfortunately, a number of lessons were learned from the *Exxon Valdez* oil spill in 1989—the worst oil spill in U.S. history. Global media coverage showed the world the desecration of a once-pristine, natural ecosystem due to a string of poor decisions by a large corporation. When the *Exxon Valdez* tanker ran aground in Prince William Sound, Alaska, it spilled nearly 11 million gallons of crude oil into the Alaskan waters, impacting over 1,100 miles of shoreline in ensuing years, hurting ecosystems and injuring and/or killing much wildlife, and depressing the economy of the local communities. The cleanup took three years and cost over $2.1 billion, not including the millions of dollars involved in legal fees and penalties. Today, scientists are still monitoring the ecosystem, testing for long-term impact and recovery. Because of the accident, environmental regulations have tightened for many industries, and tankers are now built with double hulls—if one layer ruptures, the internal layer protects the oil from spilling.

Since the 1990s, companies have started taking active steps to address environmental issues in their business operations. Authors David Levy and Charles Jones, in their article "Business strategies and climate change, United States" (Encyclopedia of Earth; http://www.eoearth.org; November 20, 2008), cite that "business, especially large multinational corporations, have de facto become a key part of the fabric of global environmental governance. In their role as investors, polluters, innovators, experts, manufacturers, lobbyists, and employers, corporations are central players in environmental issues. The recognition by governments and non-governmental organizations that large firms are not just polluters, but also possess the organizational, technological, and financial resources to address environmental problems, has stimulated consideration of ways to harness and direct these resources toward desirable goals."

The article goes on to point out how corporate climate strategy is evolving into a standard part of business for many organizations. According to a report by Ceres (a national network of investors and environmental organizations), companies are responding positively to climate change by taking such steps as establishing climate-change task forces and incorporating climate change into strategic planning processes.

THE JOB

Every year, more companies are taking proactive steps to reduce their carbon footprint. And while there are still some organizations out there that are being forced to change their business practices to comply with environmental regulations, many others are strategizing first—to prevent problems and penalties, and to improve their relationships with communities, enhance their image, and boost their bottom line by taking such steps as creating corporate climate strategy and environmental strategy task forces.

Companies hire corporate climate strategists to help them create clear strategies that address climate change. Strategists do a great deal of research before setting goals that help companies improve

(continues on page 20)

 **INTERVIEW**

Neal Reardon, Sustainable Growth Analyst

Q. What is corporate climate strategy?

A. Corporate climate strategy, simply put, is how a firm addresses the risks and opportunities presented by climate change. An effective corporate climate strategy should be proactive, clearly stated, and tailored to the firm's strengths. Rather than being an obligation or liability, the climate strategy department should seek to add value. Alternatively stated, instead of just trying to minimize damage, an excellent climate strategist maximizes benefits.

Creating a firm's climate strategy will require difficult choices as no one firm can do everything. Because of this, corporations of the future will not only define their business niche, but their climate strategy niche as well. Failing to create a clear strategy will leave firms subject to risks related to climate change, as well as constant criticism of outside stakeholders.

Q. When did companies become interested in climate strategy as part of their business practice?

A. Some companies have always had a conservation ethic, but I would say climate strategy entered the mainstream in the past five years. Walmart's sustainability initiative was a major signal that corporate social responsibility (CSR) and climate strategy were no longer the realm of fringe players. [You can find the Walmart Sustainability Report at http://walmartstores.com/Sustainability/7951.aspx.]

Q. What does a corporate climate strategist do?

A. • Creates goals—These should be clear and measurable (e.g., reduce carbon intensity of firm revenues from 10kg CO_2e per dollar earned today to 5kg CO_2e per dollar earned by 2011). The most effective goals are also synergistic with company culture.

• Asks for input—Elicits suggestions from as many departments (distribution, production, sourcing, etc.)

as possible for improving environmental performance. This is important for two reasons: It encourages a variety of ideas from many sources, and it gets employees onboard.

- Communicates at the C-level [CEO level]—Ensures that climate strategy is not pigeonholed in a corner office, but instead is an integral part of the firm. By encouraging management to consider environmental effects of business decisions from the get-go, the climate strategist avoids being caught in a game of constant damage control. The best ways to achieve this are through regular dialogue, publicly stated environmental goals (which ensure transparency and serve as reminders), and finally by pricing environmental assets (GHGs, local air pollutants, water, etc.). I can't overstate the importance of assigning firm-wide prices, because this removes subjectivity in decision making and ties environmental performance to the bottom line.

- Monitors, evaluates, and publicly communicates the climate change strategy—As I said, a firm's climate strategy should be proactive and based on clearly stated goals, but it should also be dynamic. At certain regular intervals, you should stop and take a step back to reassess your stated goals and how your performance compares to them. If something isn't working, how can it be improved? Climate strategy is a fairly new field, and there is absolutely no shame in constantly reevaluating your personal and company-wide performance. Think of climate strategy, and CSR, as a journey, not a destination.

Q. What do you do in your job? What are the day-to-day tasks?

A. • Financial and environmental analysis of business decisions (i.e., "If we used 100 percent biodiesel in all of our trucks, how many tons of GHGs would we reduce, and would it cost more or less than business-as-usual?")

(continues)

(continued)

- Present options to company management and seek feedback
- Independent data collection: academic journals, technology-specific specifications, and reviews of public renewable energy projects
- Send out requests for information (RFIs) to technology vendors
- Meet with third-party consultants regarding implementation and logistics
- Speak with government agencies regarding laws, policies, and land-use rights
- Speak with industry colleagues about their sustainability initiatives
- Create action plans and business models around selected climate mitigation investments
- Speak at public events and industry conferences

Q. What do you like most about your work?

A. My work is interesting, challenging, and gives me a personal sense of satisfaction when milestones are met. I have excellent exposure to VPs as well as CEO/CFOs, and like that I am charged with a large deal of responsibility and independence. I am exposed to a great deal and continue to learn about exciting developments in technologies, policies, and business strategies related to climate change. Attending conferences and hearing about how others are addressing climate-change risk and CSR on a practical level is thought provoking and fun.

Q. And what do you like least?

A. Making the strictly business case for climate strategy is sometimes based on things outside of my control. For example, one proposal I worked on for a [long time] was extremely cost effective when gas prices were at $4 per gallon during summer of '08, but that same proposal was much less attractive when gas prices plummeted in the following months. Also, having to repeatedly defend against criticisms from stakeholders' groups who had good intentions but poor information was at times discouraging.

Q. What was the most surprising thing about the job?

A. Everyone at the firm, from logistics to marketing, seemed sincerely interested in my job and what I was doing. My supervisor, who performed the firm's carbon footprint, gained an in-depth understanding of all aspects of the business that most managers didn't have while collecting data for her analysis. This allowed her to position herself as an excellent candidate for promotion to upper management level and she was promoted to director last year!

Q. What types of companies hire corporate climate strategists?

A. Most forward-thinking companies today are hiring climate strategists, and these companies recognize that failing to address climate change will put them at a competitive disadvantage in the future. Insurance, energy, transportation, real estate, consumer products, finance, and government are all large hubs for strategists.

Q. What does the future look like for this career?

A. This career is becoming more mainstream every day. Climate change strategies for businesses are no longer about simply generating appeal to the lifestyles of health and sustainability (LOHAS) consumer group. Now they are about making operations efficient from a sourcing, production, distribution, and post-consumption perspective.

By examining operations through the lens of climate change mitigation, cost-savings and other business opportunities often arise. For example, my work in analyzing renewable energy investments for a consumer products company lead our firm to consider entering the energy business to take advantage of economies of scale. This is a business opportunity that would never have emerged under a traditional framework.

(continued from page 15)

the environment and community. They also may recommend new business practices and technologies to help companies meet these goals. Strategists study the company's culture, solicit ideas from employees, and attend conferences about sustainability.

Neal Reardon, former sustainable growth analyst for FIJI Water, gives a better idea of what it takes to be a corporate climate strategist in the preceding interview. Reardon is currently working for the California Public Utilities Commission, based in San Francisco, as part of the team that administers the California Solar Initiative. He holds a bachelor's degree in environmental science, policy and management from the University of California at Berkeley, and a dual MBA/master of environmental policy degree from the Monterey Institute of International Studies.

REQUIREMENTS
High School
Corporate climate strategists need strong research, writing, and communication skills to generate reports and share ideas and suggestions with clients. Be sure to take courses in English, writing, and speech. Course work in science, especially environmental science, as well as math and computers will also help provide a solid foundation for this type of work.

Postsecondary Training
Corporate climate strategists have diverse educational backgrounds. "The good news is that there are many paths for addressing climate strategy," says Neal Reardon. "In fact, having team members from a variety of backgrounds is an asset in climate strategy, just like it is in traditional consulting teams. I've met people with backgrounds in engineering, law, science, operations, sourcing, and product management."

The field is still new, so be sure to find out if colleges and universities you're interested in attending offer programs in climate or environmental strategy. A bachelor's degree is recommended as a solid foundation for this career. Course work in business, management, communications, environmental issues and policy, math, English, and computers is useful.

Those with advanced degrees may find more opportunities for work. Reardon chose a graduate degree program that would specifically position him in the business/environmental performance nexus. He attended the Monterey Institute of International Studies (http://www.miis.edu), which offers a dual degree in international business (MBA) and international environmental policy (MA). He specialized in renewable energy, and was the first student in the school's history to complete the three-year program in two years. Reardon says, "In addition to traditional business and policy course work, this program integrated global trade, conflict resolution, languages, and developing economies. Studying with students from around the world also provided a wide range of perspectives on traditional business challenges."

Other Requirements

Successful corporate climate strategists need to be up to date on environmental policy, climate strategy trends, and what competitors are doing to address climate change issues. According to Neal Reardon, a broad familiarity with developments in the scientific, policy, business, and technology realms is most helpful. He says, "An effective climate strategist should know the physical drivers of climate change, as well as relative policies at the global, national, and state levels, as well as emerging development in climate mitigation and clean technologies."

Strong analytical and communication skills, and the ability to work well with other people are also essential in this type of work.

EXPLORING

To learn more about corporate climate strategy and sustainability issues, join professional associations such as Net Impact (http://www.netimpact.org), American Solar Energy Society (http://www.ases.org), and the National Association for Environmental Management (http://www.naem.org). Be sure to check out such Web sites as http://www.greentechmedia.com, http://www.grist.org, and http://alternative-energy-news.com. It's helpful to read company-specific blogs—such as Google, Sun, and Fijigreen—related to reducing the footprint of greenhouse gas emissions. Two must-read books are *The Green Collar Economy*, by Van Jones (HarperOne: New York, N.Y., 2008), and *The Truth about Green Business*, by Gil Friend (FT Press: Saddle River, N.J., 2009).

EMPLOYERS

Corporate climate strategists work in such industries as consumer products, energy, insurance, real estate, and finance, as well as for government agencies. In 2006 there were about 678,000 management analysts employed in the United States. The types of companies that are developing corporate climate strategies include FIJI Water, Xerox, DuPont, Alcoa, GE, Burt's Bees, Interface, Adidas, BP, Pepsi, and Coca-Cola Enterprises.

STARTING OUT

Volunteering is an excellent way to learn more about a company's culture, its employees, environmental issues, and to see if the work they do interests you. You may also find it a great way to meet key players, which could help you get a foot in the door. Visit the Web sites of the companies that interest you to see if they list volunteer opportunities. If you're interested in the energy-generation sector, check out GRID Alternatives (http://www.gridalternatives.org). "They offer free training sessions for volunteers who then work together to install solar panels," Reardon says. "These types of volunteer activities are a great way to learn, and should be fun!"

Attending conferences is another way to learn what industry leaders are discussing and to find out about potential employment opportunities. Visit Green Power Conferences (http://www.greenpowerconferences.com) to learn about upcoming events.

ADVANCEMENT

The career path for corporate climate strategists is not linear, so advancement can be in any number of directions, depending on one's interests and talents. Strategists can advance to senior management and director positions within the corporation. They may consult with outside organizations while employed, or start their own consulting businesses. They may also lecture at conferences, teach at universities, or share their knowledge by writing books and articles.

EARNINGS

Salary ranges are tough to locate for corporate climate strategists because the field is still relatively new. The closest job title the U.S. Department of Labor shows earnings for is management analyst. In 2008 management analysts had median incomes of $73,570, with the lowest 10 percent earning $41,910 or less and the highest earners bringing home $133,850 or more per year. You might also be able to find salary ranges at http://www.glassdoor.com, an employment Web site that features salary comparisons by job and company, employee reviews of jobs and companies, as well as actual interview questions and answers.

WORK ENVIRONMENT

Corporate climate strategists usually work in business offices, putting in at least 40 hours per week during business hours (typically Mondays through Fridays). They may need to travel on occasion to meet with employees of other corporate branches, as well as to attend conferences and events.

OUTLOOK

The proliferation of new and more stringent environmental regulations is causing more companies to seek help from corporate climate strategists, so the employment outlook is bright. More job opportunities are also expected to develop as a result of the Obama administration's Recovery and Reinvestment Act (2009), which includes more than $60 billion in clean-energy investments, and aims to invest over $150 billion over the next decade to clean-energy research and development. With the administration's plan to protect American consumers by "closing the carbon loophole" and "cracking down on polluters," corporations will be rethinking their business operations and consulting with strategists for advice on environmental compliance and corporate social responsibility.

FOR MORE INFORMATION

Learn more about corporate climate strategy news, events, and publications by visiting Ceres' Web site.

Ceres
99 Chauncy Street, 6th Floor
Boston, MA 02111-1703
Tel: 617-247-0700
http://www.ceres.org

This program aims to reduce greenhouse gas emissions by setting standards and tracking projects through an online registry.

Climate Action Reserve
523 West Sixth Street, Suite 428
Los Angeles, CA 90014-1208
Tel: 213-891-1444
http://www.climateactionreserve.org

Set up by the World Meteorological Organization and the United Nations Environment Programme, IPCC offers objective information about climate change.

Intergovernmental Panel on Climate Change (IPCC)
Tel: 41-22-730-8208
Email: IPCC-Media@wmo.int
http://www.ipcc.ch

Learn more about the Energy & Environment section of the American Recovery and Reinvestment Act by visiting the White House Web site.

The White House
1600 Pennsylvania Avenue, NW
Washington, DC 20500-0003
http://www.whitehouse.gov/issues/energy_and_environment/

Eco-investors

OVERVIEW

Eco-investors help finance new and existing businesses that promote conservation, prevent global warming, and enhance communities and the environment. They often team up with other individuals, financial groups, and financial advisers for guidance in funding green businesses appropriate to their investment principles and missions.

HISTORY

From the late 1800s through the 1950s most investors in private companies were wealthy families such as the Rockefellers, Vanderbilts, and Morgans. The Rockefellers funded numerous companies and helped start Eastern Air Lines and Douglas Aircraft. The Vanderbilts, whose wealth stemmed from the railroad and shipping empires Cornelius Vanderbilt built in the previous century, invested in universities. The Morgans, who rose to fame through their success in turning troubled railroads around in the late 19th century, invested in new technologies and engineering projects, including the

Brooklyn Bridge, the base of the Statue of Liberty, and the Panama Canal.

After World War II, private equity firms developed, making it possible for people other than the wealthy to invest in business ventures. The American Research and Development Corporation (ARDC) and J.H. Whitney & Company, each founded in 1946, were the first two venture capital firms in the United States.

ARDC was the brainchild of Georges Doriot, a former Harvard Business School dean who was considered the "father of venture capitalism." He, along with Ralph Flanders and Karl Compton, created ARDC to encourage people to invest in businesses run by soldiers who had returned from the war. The first venture capital success is attributed to ARDC: In 1957, it invested $70,000 in the Digital Equipment Corporation (DEC), and by 1958, just one year later, DEC was valued at more than $355 million. ARDC invested in numerous companies until Doriot's retirement in 1971, when it was merged with Textron, a company that still exists. Doriot noted, at the end of his career, that the next generation of venture capitalists did not have the same interest, patience, and solicitude as his generation had regarding the types of businesses they funded. He felt that venture capitalism had become speculative—investors were not focused on the intention of the business, but rather on the money they were going to make. He was interested in the quality of the business idea, its mission, its ethics—a mindset that closely resembles that of eco-investors today.

John Hay Whitney was a savvy investor who founded Pioneer Pictures in 1933 and acquired 15 percent of the shares in Technicolor Corporation. He and his partner Benno Schmidt founded J.H. Whitney & Company (1946), which continues to invest in companies today. Whitney's most famous venture was his investment in Florida Foods Corporation, which developed a way to concentrate orange juice into a powder, thereby enabling wider distribution to American soldiers at war. The company created Minute Maid— frozen concentrated orange juice for the public. The Coca-Cola Company purchased Minute Maid in 1960.

Investors in the 1960s and 1970s focused on starting and expanding companies, predominantly those that were developing revolutionary electronic, medical, and data-processing technologies. Also in the early 1970s, environmental legislation was introduced that inspired people to create businesses with socially responsible missions. It was during this time also that private equity firms set up an operating structure that is still followed. They organized limited partnerships that held the investments, and the general partners and investors put up the capital.

Since then venture capital has had its ups and downs. By the mid-1980s, there were more than 650 venture capital investment firms in the United States; however, the 1987 stock market crash cooled the investment field. Scarred by the crash, people thought it safer to put their money into older, established companies with stable histories. Things picked up again in the late '90s, when there was a surge of interest in all things to do with the Internet and computer technologies (known as the dot.com days). And then the NASDAQ crash and technology slump in 2000 forced many venture capital firms to reduce or drop their investments entirely.

Despite a recent economic slowdown, venture capitalists are still interested in funding businesses. According to a MoneyTree report by PriceWaterhouseCoopers and National Venture Corporations Association, based on data by Thomson Reuters, venture capitalists invested $28.3 billion in 3,808 business deals in 2008. While this was a decline from previous years, the Clean Technology sector was the only sector to achieve significant growth: $4.1 billion invested in 277 deals in 2008, averaging to a 52 percent growth in dollars when compared to 2007. Clean Technology encompasses alternative energy, pollution and recycling, and power supplies and conservation.

THE JOB

Eco-investors use their money to invest in companies that help the environment. They often help finance new business ventures as well as currently existing green companies. The types of companies they help fund usually do such things as provide clean energy technologies and solutions, or offer products and services related to solar and wind energy, recycling, waste management, and carbon-emission reduction.

Eco-investors can be individuals, foundations, or groups of individuals who pool their money together to fund green businesses that have high growth potential. Eco-investors may also be known as *venture capitalists* or even *green angels*. They usually invest cash in exchange for shares in the company's business. Most small green companies and start-ups have difficulty securing bank loans because their businesses don't have a proven track record. Those entrepreneurs that have innovative ideas, strong business models, technical and industry expertise, and solid management teams have better odds of attracting venture capitalists.

Eco-investors read business plans, income statements, balance sheets, and shareholder reports to gain a better understanding of the business idea and its potential in the marketplace. Small business owners may also pitch their ideas to investors through presentations, which can be in person as well as at new-business pitch conferences and business site tours. Eco-investors spend a great deal of time researching these businesses, industry trends, and growth potential—consulting with financial advisers and analysts, and other experts, throughout the process—to decide the feasibility of investing. For the hundreds of proposals that cross their desks, they usually approve only a handful for funding.

Venture capital funding is offered for different stages of business, depending on where the company is in its development. The stages are as follows:

- Seed money, to finance a new idea
- Start-up, for marketing and product development costs
- First-round, to cover sales and manufacturing expenses
- Second-round, usually for new companies that are not yet turning a profit
- Third-round (mezzanine), for a company that's just started profiting to expand its business
- Fourth-round (bridge financing), to help companies "go public"

What types of businesses might an eco-investor go for? The Investors' Circle (IC) (http://www.investorscircle.net), a network of socially responsible investors based in San Francisco, cites that their members lean toward investing in energy and environment; food and organics; education and media; health and wellness; and community and international development. A specific example of a green business that Investors' Circle backed is New Jersey-based TerraCycle. Started in 2001, their original product was worm excrement packaged in recycled soda bottles and sold as fertilizer. Since then, TerraCycle has branched out into other garden-related products, such as seed starters, pots, repellants, cleaners, and bird feeders, all made from recycled materials. Other IC-funded companies include Organic To Go, an organic food retail chain, and Virgin Money USA (now owned by Richard Branson, the founder of Virgin).

REQUIREMENTS
High School
Take classes in math, business, marketing, and science if you are interested in this type of work. A large part of the job entails reading business proposals and discussing business strategies with financial advisers and green entrepreneurs, so having strong verbal and written communication skills is an asset. English, writing, and speech classes will help provide a good foundation.

Postsecondary Training
Eco-investors have diverse educational backgrounds. A college degree is not required, but most people in this field have a bachelor's degree in business. Course work in finance, economics, business, marketing, accounting, and statistics is recommended. Classes in real estate, land development, science, sustainable business practices, and environmental regulations and policies are also helpful. Many eco-investors have a master's degree in business, which can provide a deeper understanding of business practices and attract potential employers.

Other Requirements
Venture capitalism can be a risky business. There is no guarantee that a company will succeed, and when a company fails, the investor loses money. Risk-takers, adventurers, and open-minded business explorers are happiest in this type of work. A passion for helping communities thrive and for improving and enhancing the environment is also a key driving force behind eco-investing. Investors read and research a great deal, so a willingness to continually learn and probe deeper is essential.

EXPLORING
Read and learn everything you can about green issues, trends, and technology by regularly visiting the Web sites of such organizations as Social Funds (http://www.socialfunds.com) and Environmental Leader (http://environmentalleader.com). Keep up with environmental investing by reading magazines such as *Seed* (http://seedmagazine.com) and *Environmental Finance* (http://www.environmental-finance.com). You can also learn more about environmental banking by checking out the Environmental Bankers Association (http://www.envirobank.org).

EMPLOYERS

Eco-investors may own their own businesses or work on staff for venture capitalist firms, banks, finance companies, and other finance-related businesses. Some inherit money that they use to fund businesses. A small sample of environmental venture capitalist firms that hire eco-investors includes Environmental Capital Partners (http://www.ecpcapital.com), which is affiliated with the New York Private Bank & Trust; Expansion Capital Partners (http://www.expansioncapital.com); and Foundation Capital (http://www.foundationcapital.com). You can learn about many other similar companies by using a search engine such as Google and plugging in the key search words "environmental venture capitalist."

STARTING OUT

Eco-investors take many paths in their careers. Many start out by working for banks, financial firms, securities dealers, insurance companies, and other finance-related organizations. The best way to learn firsthand about eco-investing is to intern or work part time at a venture capitalist firm. Use the Internet to find organizations near you, and then contact them to see if they offer any internship programs or need part-time office help. Getting a foot in the door and making connections is the first step in learning if this career suits you.

ADVANCEMENT

Eco-investors who work for venture capital firms can advance by taking on more responsibilities and moving up to positions of greater authority. Some eco-investors may have gotten their start in the industry while at a venture capital firm, and can advance by starting their own venture capital company. Teaching at universities, writing for various media, and speaking at professional conferences are other ways to expand skills in the eco-investment field.

EARNINGS

Eco-investors' salaries vary widely because they have diverse work backgrounds. Those who work as financial managers had median annual incomes of $99,330 in 2008, with the lowest 10 percent earning $53,680 and the highest 10 percent earning $166,400 or more, according to the U.S. Department of Labor. Personal financial advisers had slightly lower median incomes of $69,050 in 2008.

Environmental bankers who work for large institutions will receive higher salaries. Bankers may also receive bonuses and stock options, as well as tuition reimbursement, health insurance coverage, and paid vacation and sick time.

WORK ENVIRONMENT

Eco-investors usually work during normal business hours in offices or in their homes. They spend some time traveling to meet business owners and look at business sites. They also keep up with environmental investment trends and issues by attending lectures, workshops, and conferences.

OUTLOOK

Investing in green business is still a relatively new field, but the need for eco-investors should increase in the years to come. The growing population, changes in the environment, and rising energy costs are creating greater demand for businesses that offer solutions. New environmental laws and regulations are inspiring more people to come up with business ideas that reduce carbon emissions, pollution, and waste. And these businesses, as well as current companies needing to change their operations to meet new parameters, will need funding. According to the U.S. Small Business Administration, approximately 600,000 new businesses are started each year in the United States. The percentage of those that are green businesses is not broken down, but it would be safe to say that a number of them are sustainable ventures requiring venture capital.

FOR MORE INFORMATION

For articles, publications, and upcoming events about eco-investing and green entrepreneurship, visit

Eco Investment Club
The Pannikin Building
655 G Street, Suite E (Blik Design Studios)
San Diego, CA 92101-7039
Tel: 866-960-9495
http://www.ecoinvestmentclub.com

For more information about green banking, contact

Environmental Bankers Association (EBA)
510 King Street, Suite 410
Alexandria, VA 22314-3132
Tel: 703-549-0977
Email: eba@envirobank.org
http://www.envirobank.org

Learn more about venture capital industry standards, news, and membership by visiting NVCA's Web site.

National Venture Capital Association (NVCA)
1655 North Fort Myer Drive, Suite 850
Arlington, Virginia 22209-3199
Tel: 703-524-2549
http://www.nvca.org

For information about socially responsible investing, contact

Social Funds
SRI World Group, Inc.
74 Cotton Mill Hill, A255
Brattleboro, VT 05301-7808
http://www.socialfunds.com

Environmental Bankers

OVERVIEW

Environmental bankers help customers and clients invest their money in ways that enhance their homes, businesses, communities, and the environment. Entrepreneurs with green businesses look to environmental bankers for loans, extensions of credit, and for investment and money management advice.

HISTORY

The U.S. Congress established the first central Bank of the United States in 1791, which lasted until its charter expired in 1811. The second bank, by the same name, existed from 1816 to 1832. In those early days, lending practices were strictly regulated and bankers were careful about loaning money to businesses. Long-term loans were not considered. Typically, business owners and manufacturers who needed money to pay their workers and suppliers could get a short-term loan from the bank. They were expected to repay it within 30 to 60 days. Farmers could also borrow money for equipment and farm product shipments.

From 1832 to 1864 state governments took over the banking system; this was a confusing and frustrating time for bank customers. Each bank had its own currency: bank notes, which were supposed to be convertible to gold or silver. People would bring their notes to the bank, and it was up to the banks to verify that the notes were valid. But back then, banks frequently did not have enough cash to redeem the notes. Also, because of the overwhelming variety of notes in existence (10,000 different types were circulating in this country by 1860), it was difficult to confirm that they were real. In the end, many people found themselves holding pieces of paper that had no value. To end the confusion, the National Currency Act (1863) and the National Bank Act (1864) organized currency and created tighter regulations and regular banking examinations through the then-new Office of the Comptroller of the Currency (OCC).

New banking laws and regulations were passed in 1933 to address the banking depression (1929–1933), which occurred when banks had failed and defaulted on people's loans. To rebuild confidence in the banking system, Congress enacted federal deposit insurance, and President Franklin D. Roosevelt tasked the OCC to examine all banks and either allow them to reopen or liquidate them. The OCC continues to this day to oversee national banks and enforce banking laws.

Since the 1970s, when environmental legislation was initiated, concern for the environment has grown, and environmental banks have since evolved. Within the past two decades, the number of environmental banks in the United States has multiplied nearly 10 times over, rising from 46 in 1993 to over 400 banks nationwide by 2000, and the numbers continue to rise.

THE JOB

Environmental bankers work for banks and financial institutions that focus on promoting and financing green initiatives. Most banks consider green business start-ups to be risky and are therefore reluctant to help finance them. Green banks, however, will work with businesses that have missions regarding sustainability and that also fill a certain niche in the marketplace. These emerging businesses may include organic markets and restaurants, solar or wind power companies, eco-resorts, eco-developers, and more.

A green bank is also more than just one that offers online ("paperless") banking. Many banks are promoting themselves as green because of this tactic. And while saving trees is certainly impor-

Dave Williams, president of ShoreBank Pacific, left, and Bonnie Anderson, vice president of real estate, stand outside a green residential development that the financial institution is helping to fund. *AP Photo/The Daily Journal of Commerce, Dan Carter*

tant in preserving the environment, this sole activity is not the full definition of green. A real green bank does much more. It carefully examines the businesses it funds and makes sure their missions and business operations benefit the environment and community. To do this, a growing number of banks are creating environmental banking groups to research and finance environmentally friendly activities and help prevent global warming.

Another type of environmental banking concerns permanently reserving public or private land for protection of a critical habitat. These are known as mitigation banks and can be categorized as a wetland, upland, or species bank.

"Environmental banking is a form of environmental protection that is also a business opportunity for the private sector," according to Critical Habitats, an environmental banking company based in Denver, Colorado. It describes environmental banking as the "creation and sale of a natural resource commodity or a pollution reduction."

In the past, developers and environmentalists clashed over the development of areas where species and/or plant life was threatened. Since environmental legislation was introduced in the 1970s, local, state, and federal authorities (e.g., U.S. Army Corps of Engineers, Environmental Protection Agency, etc.) review and approve or reject developments that impact critical habitats. The laws have dictated that developers re-create the ecosystems or habitats that are destroyed during development. This re-creation is called onsite mitigation and can often add years to the construction process. Developers also had to pay fines for the damage, but government agencies had trouble enforcing the laws and collection payments. Environmental bankers have since come along to act as third parties; they offer developers the option of buying a "credit" instead.

A credit is usually one acre of restored critical habitat, or it can be a specific number of a member of an endangered species. An environmental bank is formed when a landowner, investor, or municipality buys a feasible tract of land, makes it habitable for plants and animals, and then sells credits to this land to developers who need it to compensate for damage they've caused during development. After developers buy the credits, they are released from permit responsibilities and can continue with their project. Increased road and highway construction has the potential to damage wetlands and other ecosystems. State departments of transportation realized early on the benefits of offsetting habitat damage through mitigation banking (the restoration, creation, enhancement, or preservation of a wetland, stream, or habitat conservation area which offsets expected adverse imputs to similar nearby ecosystems).

Environmental bankers help people find appropriate land for environmental banks. They meet with clients to discuss their needs and then conduct feasibility studies to determine the potential for critical habitat creation. They help secure appropriate government approvals and certification, and also help with financing and selling credits.

REQUIREMENTS
High School
Course work in business and math is an essential foundation for this job. Understanding environmental issues is important, so be sure to take science classes as well. Environmental bankers meet regularly with clients to discuss their needs and advise them on the best course of action to take; English and writing classes will be helpful in this arena.

Postsecondary Training
Most financial institutions require financial managers and financial advisers to have a bachelor's degree in finance, business administra-

tion, accounting, statistics, or economics. Undergraduate students usually take classes in statistics, economics, and business, as well as corporate budgeting and financial analysis. Course work in sustainable business practices and environmental regulations and policies is also recommended, as are science classes. Students pursuing financial adviser careers take classes in investments, taxes, risk management, land development, and estate planning. Those with a graduate degree may have greater work opportunities, particularly if they have knowledge of risk management.

Certification or Licensing

Licensing or certification requirements will vary depending on the company the environmental banker works for and the type of work he or she does. To get a financial license, employees must first get sponsorship from the company they work for. And when they change companies, they need to renew their licenses. In the securities industry, the Financial Industry Regulatory Authority is the main licensing organization.

Environmental bankers who work as personal financial advisers need the Series 7 and Series 63 or 66 licenses. These licenses give them the right to act as a registered representative of a securities firm and to give financial advice. The Series 7 license requires company sponsorship, which means that self-employed personal financial advisers have to maintain a relationship with a large securities firm. This relationship will enable them to represent that firm in the buying and selling of securities. To sell insurance, financial advisers need additional licenses issued by state licensing boards.

Voluntary certification can also help advance the careers of financial analysts and advisers. Analysts with a bachelor's degree, four years of work experience, and who have passed three exams can receive the chartered financial analyst designation from the CFA Institute. Advisers can earn the certified financial planner credential from the Certified Financial Planner Board of Standards if they have three years of relevant work experience, a bachelor's degree, and also pass an exam.

Other Requirements

To succeed in this work, environmental bankers must have strong math and analytical skills, and the ability to solve problems. To help clients understand complicated financial strategies, bankers must be organized, detail-oriented, clear communicators. Those who run their own businesses must be self-motivated and reliable. A deep understanding of the environment, green business practices, land development, and local and national environmental policies is also essential.

EXPLORING

Learn more about environmental banking by visiting the Web sites of such organizations as the Environmental Bankers Association (http://www.envirobank.org) and Environmental Finance (http://www.environmental-finance.com). Use the Internet to search for green banks, and then see if you can contact an environmental banker to find out what they do, and what they like most and least about their job. Keep up with financial news in general by reading magazines like *Forbes*, *Fortune*, and *Money*. And stay current on environmental issues by reading *Audubon* (http://www.audubonmagazine.org) and *Sierra* (http://www.sierraclub.org/sierra) magazines.

EMPLOYERS

Environmental bankers work on staff for banks, finance companies, and other finance-related businesses. They also work for environmental firms such as Critical Habitats. Some are hired as consultants to companies that are transforming their businesses to sustainable operations. Environmental bankers may work for such institutions as the New Resource Bank of San Francisco, California, and the ShoreBank Corporation, based in Chicago, Illinois, and with offices throughout the United States.

In 2006 about 506,000 financial managers were employed in the United States. More than 30 percent worked for banks, savings institutions, finance companies, credit unions, insurance carriers, and securities dealers. And about 8 percent worked for the federal, state, or local government. In that same year, approximately 221,000 financial analysts and 176,000 financial advisers were employed throughout the country. About 30 percent of the advisers were self-employed, operating their businesses predominately in urban areas.

STARTING OUT

Working as an intern or on a part-time basis at a green bank is a good way to get a foot in the door and learn if this field is for you. Use a search engine such as Google to see if there are green banks near you. You can also learn about environmental jobs and companies by visiting http://www.sustainablebusiness.com and http://www.greenjobs.com.

ADVANCEMENT

Environmental bankers who run their own businesses can advance by adding more clients to their roster and managing more assets. Those who work for firms may advance by taking on more responsibilities, working on more important projects, and moving into management positions. They may also advance by leaving the firm and starting their own business. Another way environmental bankers can advance is by increasing their knowledge about specific issues and topics, either through workshops, certification, or advanced degrees.

EARNINGS

Environmental bankers' salaries will vary depending on the type of work they do and their employer. Environmental banking is not specifically addressed in the Bureau of Labor Statistics, although an idea of the salary range for those working in the financial sector can be drawn from the earnings reported for financial advisers and financial managers. According to the U.S. Department of Labor, personal financial advisers had annual median salaries of $69,050 in 2008; the lowest 10 percent earned $34,390, and the highest 10 percent earned $166,400 or higher. Financial managers had higher salaries, with the lowest 10 percent earning $53,680, the middle 50 percent averaging $99,330, and the highest 10 percent earning a minimum of $166,400.

Environmental bankers who work for large institutions will receive higher salaries. Bankers may also receive bonuses, stock options, tuition reimbursement, health insurance coverage, and paid vacation and sick time.

WORK ENVIRONMENT

Environmental bankers usually work in offices or in their homes. Most work at least 40 hours per week or more. They spend some time traveling for client and investor meetings, and to attend business conferences, training sessions, and workshops. Those who teach and lecture spend evenings or weekends at other organizations and institutions. And those who work for environmental banks that provide mitigation bank services will travel to look at sites and meet with clients.

OUTLOOK

Employment of financial managers overall is expected to grow about as fast as the average for all occupations through 2016, according to the U.S. Department of Labor. Expansion of the global economy and an increase in regulatory reforms will create more jobs for those with financial expertise. The interest in investing in companies that benefit communities and the environment and the need for banks that offer green investing options will likewise grow. According to a Javelin Strategy & Research study in November 2008, approximately 43 percent of consumers polled said they would prefer to do business with green banks.

Mitigation banking is also on the rise, according to findings from a survey by the Environmental Law Institute. Only 46 mitigation-banking sites existed in 1992, with 40 percent of those located in California or Florida. By 2001 the practice of mitigation banking had spread to 42 states and included a total of nearly 350 projects.

Competition is usually strong for financial positions, especially during times of economic slowdowns. Environmental bankers with advanced degrees and/or certification in specialized areas of finance and financial management will have the edge in the job market.

FOR MORE INFORMATION

For financial adviser career information, contact
Certified Financial Planner Board of Standards Inc.
1425 K Street, NW, Suite 500
Washington, DC 20005-3686
Tel: 202-379-2200
http://www.cfp.net

For information on financial analyst careers, contact
CFA Institute
PO Box 3668, 560 Ray C. Hunt Drive
Charlottesville, VA 22903-2981
Tel: 800-247-8132
Email: info@cfainstitute.org
http://www.cfainstitute.org

For more information about critical habitats and environmental banking, visit
Critical Habitats Inc.
4486 South Wolff Street

Denver, CO 80236-3327
Tel: 303-679-8262
Email: communicate@criticalhabitats.com
http://criticalhabitats.com

For news and articles about environmental banking, and to learn more about career opportunities in the field, visit EBA's Web site.

Environmental Bankers Association (EBA)
510 King Street, Suite 410
Alexandria, VA 22314-3132
Tel: 703-549-0977
Email: eba@envirobank.org
http://www.envirobank.org

For information about securities industries employment, contact

Financial Industry Regulatory Authority
1735 K Street, NW
Washington DC, 20006-1506
Tel: 301-590-6500
http://www.finra.org

Environmental Economists

QUICK FACTS

School Subjects
Business
Economics
Mathematics

Personal Skills
Helping/teaching
Technical/scientific

Work Environment
Primarily indoors
Primarily one location

Minimum Education Level
Master's degree

Salary Range
$44,050 to $83,590 to
$149,110+

Certification or Licensing
None available

Outlook
As fast as the average

OVERVIEW

Environmental economists are concerned with how society uses resources such as land, labor, raw materials, and machinery to produce goods and services for consumption and production, and what impact the use and production have on the environment. They study how economic systems address three basic questions: "What shall we produce?" "How shall we produce it?" and "For whom shall we produce it?" The economist then compiles, processes, and interprets the answers to these questions. There are about 15,000 economists employed in the United States.

HISTORY

Economics deals with the struggle to divide up a finite amount of goods and services to satisfy an unlimited amount of human needs and desires. No society, no matter how rich and successful, is able to produce everything needed or wanted by individuals. This reality was evident to people throughout history. In ancient Greece, the philosopher Plato discussed economic topics in his work *The Republic*, saying the division of labor among people was the only way to supply a larger need. Individuals, he said, are not naturally self-sufficient and thus they need to cooperate in their efforts and exchange goods and services.

It was not until 1776 that the theory of economics was given a name. Adam Smith, in his work *Wealth of Nations*, described that individuals, given the opportunity to trade freely, will not create chaos. Instead, he claimed that free trade results in an orderly, logical system. His belief in this free trade system has been interpreted as an endorsement of laissez-faire capitalism, which discourages government restrictions on trade. Other economists believe that regulation is necessary to limit corruption and unfair or monopolistic practices.

The importance of economics is evidenced by its status as the only social science in which a Nobel Prize is awarded. In the last century, economics has come to be used in making a broad array of decisions within businesses, government agencies, and many other kinds of organizations. And because of the mounting concern about global warming and greenhouse gas emissions, environmental economics has evolved.

THE JOB

Environmental economists grapple with many issues relating to the supply and demand of goods and services and the means by which they are produced, traded, and consumed. These specific economists focus on the economic effects of national or local environmental policies around the world. According to the National Bureau of Economic Research, a nonprofit research organization based in Cambridge, Massachusetts, environmental economists consider issues such as the costs and benefits of alternative environmental policies to deal with air pollution, water quality, toxic substances, solid waste, and global warming.

Environmental economists conduct research, collect and analyze data, monitor economic trends, and develop forecasts. They may also study such issues as energy costs, inflation, interest rates, exchange rates, business cycles, taxes, and employment levels, among others. Data gathering is an important part of all economists' jobs. They create the methods by which to reap information, such as techniques used to conduct surveys and develop forecasts. Another key task is presenting research findings in reports, using tables and charts. Economists need strong, clear writing and speaking skills; they are often sharing information with people who are not economists, so the ability to translate material in ways people understand is critical. Some economists also write for and appear in the media.

While most economists either teach at the university level or perform research for government agencies, many work for individual for-profit or not-for-profit organizations.

Economics professors teach basic macro- and microeconomics courses as well as courses on advanced topics such as economic history and labor economics. (Macroeconomics deals with the "big picture" of economics as a whole, and microeconomics deals with individual companies and persons.) They also perform research, write papers and books, and give lectures, contributing their knowledge to the advancement of the discipline.

Government economists study national economic trends and problems; their analyses often suggest possible changes in government policy to address such issues.

For-profit and not-for-profit companies both employ economists to assess connections of organizational policy to larger business conditions and economic trends. Management often will rely on this research to make financial and other kinds of decisions that affect the company.

In their education, economists usually specialize in a particular area of interest. While the specialties of university economists range across the entire discipline, other economists' expertise generally falls into one of several categories. *Financial economists* examine the relationships among money, credit, and purchasing power to develop monetary policy and forecast financial activity. *International economists* analyze foreign trade to bring about favorable trade balances and establish trade policies. *Labor economists* attempt to forecast labor trends and recommend labor policies for businesses and government entities. *Industrial economists* study the way businesses are internally organized and suggest ways to make maximum use of assets. *Environmental economists* study the relationships between economic issues and the allocation and management of natural resources. *Agricultural economists* study food production, development in rural areas, and the allocation of natural resources.

REQUIREMENTS
High School
A strong college preparatory program is necessary in high school if you wish to enter this field. Courses in other social sciences,

economics, mathematics, and English are extremely important to a would-be economist, since analyzing, interpreting, and expressing one's informed opinions about many different kinds of data are primary tasks for someone employed in this field. Also, take computer classes so that you will be able to use this research tool in college and later on. Finally, since you will be heading off to college and probably postgraduate studies, consider taking a foreign language to round out your educational background.

Postsecondary Training

A bachelor's degree with a major in economics is the minimum requirement for an entry-level position such as research assistant. Many colleges now offer degree programs in environmental and ecological economics. A master's degree, or even a Ph.D., is more commonly required for most positions as an economist.

Typically, an economics major takes at least 10 courses on various economic topics, plus two or more mathematics courses, such as statistics and calculus or algebra. Environmental economics students usually take classes in microeconomic theory, and the economics of resources and the environment. The federal government requires candidates for entry-level economist positions to have a minimum of 21 semester hours of economics and three hours of statistics, accounting, or calculus. Graduate-level courses include such specialties as advanced economic theory, econometrics, international economics, and labor economics.

Other Requirements

Economists' work is detail oriented. They do extensive research and enjoy working with abstract theories. Their research work must be precise and well documented. In addition, economists must be able to clearly explain their ideas to a range of people, including other economic experts, political leaders, and even students in a classroom.

EXPLORING

You can augment your interest in environmental economics by taking related courses in the social sciences and mathematics, and by becoming informed about business and economic trends through reading business-related publications such as newspaper business sections and business magazines. In addition to economics course work, college students can participate in specific programs and extracurricular activities sponsored by their university's business school, such as internships with government agencies and businesses and business-related clubs and organizations.

EMPLOYERS

Approximately 15,000 economists are employed in the United States. According to the U.S. Department of Labor, in 2006, the government employed 52 percent of economists, with 32 percent working in federal government and 20 percent in state and local government. Many work as researchers at government agencies such as the U.S. Department of Labor, or international organizations such as the United Nations, the World Bank, and the International Monetary Fund. Many teach at colleges and universities. Others find employment at nonprofit or for-profit organizations, helping these organizations determine how to use their resources or grow in profitability. Most economics-related positions are concentrated in large cities, such as New York, Chicago, Los Angeles, and Washington, D.C., although academic positions are spread throughout the United States.

STARTING OUT

The bulletins of the various professional economics associations are good sources of job opportunities for beginning economists. Your school's career services office can also assist you in locating internships and in setting up interviews with potential employers.

ADVANCEMENT

An economist's advancement depends on his or her training, experience, personal interests, and ambition. All specialized areas provide opportunities for promotion to jobs requiring more skill and competence. Such jobs are characterized by more administrative, research, or advisory responsibilities. Consequently, promotions are governed to a great extent by job performance in the beginning fields of work. In university-level academic positions, publishing papers and books about one's research is necessary to become tenured.

EARNINGS

Economists are among the highest-paid social scientists. According to the U.S. Department of Labor, the median salary for economists was $83,590 in 2008. The lowest paid 10 percent made less

Notable Economics Graduates

Name	Job Title	Institution
Scott Adams	Cartoonist, creator of Dilbert	Hartwick College
Jennifer Azzi	Professional basketball player	Stanford University
Steve Ballmer	CEO, Microsoft	Harvard University
Barbara Boxer	U.S. Senator	Brooklyn College
William F. Buckley	Journalist	Yale University
George H.W. Bush	41st President of the United States	Yale University
Eileen Collins	Shuttle Commander, NASA	Syracuse University
Esther Dyson	Software pioneer	Harvard University
Earl Graves	CEO, *Black Enterprise* magazine	Morgan State University
Sandra Day O'Connor	U.S. Supreme Court Justice	Stanford University
Ronald Reagan	40th President of the United States	Eureka College
Lionel Richie	Singer/songwriter	Tuskegee University
George Schultz	Former U.S. Secretary of State	Princeton University, Massachusetts Institute of Technology
Arnold Schwarzenegger	Actor, Governor of California	University of Wisconsin—Superior
John Snow	Former U.S. Secretary of the Treasury	University of Virginia
John Sweeney	President, AFL-CIO	Iona College
Ted Turner	Founder, Cable News Network	Brown University
Meg Whitman	Former CEO, eBay Technologies	Princeton University

than $44,050 and the highest paid 10 percent earned more than $149,110.

The U.S. Department of Labor reports that economists employed by the federal government earned average annual salaries of $101,020 in 2008. Starting salaries for federal government economists vary by degree attained. Economists with a bachelor's degree earned approximately $35,752 in 2007; economists with a master's degree earned approximately $43,731; and those with a Ph.D., $63,417. College economics educators earned salaries that ranged from less than $27,590 to $113,450 or more in 2006, according to the U.S. Department of Labor. Educators employed at colleges and universities had mean annual earnings of $73,207, as cited in a survey conducted by the American Association of University Professors. Four-year institutions usually pay higher salaries than two-year schools. Faculty salaries averaged $84,249 in private independent institutions in 2006.

Private-industry economists' salaries can be much higher, into the six figures. Notably, in a study published in *Money* magazine, economists' salaries tended to be 3.1 times higher at mid-career than their starting salaries. According to the survey, this is a higher increase than in any other profession; lawyers made 2.77 times more and accountants 2.21 times more in their mid-careers than at the start. Benefits such as vacation and insurance are comparable to those of workers in other fields.

WORK ENVIRONMENT

Economists generally work in offices or classrooms. The average workweek is 40 hours, although academic and business economists' schedules often can be less predictable. Economists in nonteaching positions often work alone writing reports, preparing statistical charts, and using computers, but they may also be part of a research team. Most economists work under deadline pressure and sometimes must work overtime. Regular travel may be necessary to collect data or to attend conferences or meetings.

OUTLOOK

The employment of economists is expected to grow about as fast as the average for all occupations through 2016, according to the U.S. Department of Labor. Private industry is expected to offer the most opportunities for economists with management, scientific, and consulting backgrounds. Green economy is especially growing as a field because of the concern about global warming and the complexities of environmental policies and regulations. As a result, environmental economists will be in demand in the years to come.

Openings will also occur as economists retire, transfer to other job fields, or leave the profession for other reasons. Economists employed by private industry—especially in management, scientific, and technical consulting services—will enjoy the best prospects. In the academic arena, economists with master's and doctoral degrees will face strong competition for desirable teaching jobs. The demand for secondary school economics teachers is expected to grow. Economics majors with bachelor's degrees alone will experience the greatest employment difficulty, although their analytical skills can lead to positions in related fields such as management and sales. Those who meet state certification requirements may wish to become secondary-school economics teachers, as demand for teachers in this specialty is expected to increase.

FOR MORE INFORMATION

For information on agricultural economics and a list of colleges that offer programs in the field, contact

Agricultural and Applied Economics Association
555 East Wells Street, Suite 1100
Milwaukee, WI 53202-3800
Tel: 414-918-3190
Email: info@aaea.org
http://www.aaea.org

For information on job listings and resources of interest to economists, contact

American Economic Association
2014 Broadway, Suite 305
Nashville, TN 37203-2425
Tel: 615-322-2595
Email: aeainfo@vanderbilt.edu
http://www.aeaweb.org

For information on graduate programs in environmental and resource economics, contact

Association of Environmental and Resource Economists
1616 P Street, NW, Room 600
Washington, DC 20036-1434

Tel: 202-328-5125
Email: info@aere.org
http://www.aere.org

The CEE promotes the economic education of students from kindergarten through 12th grade. It offers teacher training courses and materials. For more information, contact

Council for Economic Education (CEE)
1140 Avenue of the Americas
New York, NY 10036-5803
Tel: 800-338-1192
Email: info@councilforeconed.org
http://www.councilforeconed.org

To read the publication Careers in Business Economics, *contact or check out the following Web site:*

National Association for Business Economics
1233 20th Street, NW, Suite 505
Washington, DC 20036-2365
Tel: 202-463-6223
Email: nabe@nabe.com
http://www.nabe.com

The NBER is a nonprofit, nonpartisan economic research organization that provides publications, research, and statistics. More than half of the Nobel Prize winners in economics (16 out of 31) have been researchers at NBER.

National Bureau of Economic Research (NBER)
1500 Massachusetts Avenue
Cambridge, MA 02138-5398
Tel: 617-757-3900
http://www.nber.org

For information on membership and job listings, contact

Society of Government Economists
PO Box 77082
Washington, DC 20013-8082
http://www.sge-econ.org

Green Builders

OVERVIEW

Green builders build commercial and residential structures that have less impact on the environment and on human health than typical constructions. The buildings they help create use energy, water, and other resources efficiently and reduce waste, pollution, and environmental degradation. Depending on their job role, green builders plan, direct, coordinate, and/or budget construction and maintenance projects.

HISTORY

Green building dates back to early times when settlers chose building materials that were readily available to create structures that worked well with the environment. In the Southwest, ancient Pueblo people used mud, sand, and water to form adobe bricks that held up well to varying temperatures and climates. They also built their homes to receive solar heat in the winter.

People's interest in creating energy-efficient, resource-conserving buildings was sparked anew during the environmental movement of the 1960s and 1970s. The media helped raise awareness of environmental issues such as pollution and land degradation through such magazines as *Mother Jones* (http://www.motherjones.com) and *Environment* (http://www.environmentmagazine.org), and through books like Rachel Carson's *Silent Spring*, which pointed a spotlight at the impact of chemicals on the air, water, soil, and human health. The first Earth Day, in 1970, catapulted the environmental movement and collective consciousness to a global scale.

The energy crises of 1973 and 1977 further heightened awareness about the need to conserve energy and find renewable energy sources. America's demand for domestic oil had exceeded supply capabilities, dependence of foreign oil had strained U.S. relations with certain countries, and the federal government's attempt to regulate energy had sent oil prices soaring. Ideas were forming for sustainable building to improve quality of life, conserve resources, and save money.

The green building industry has advanced over the past 30 years and is setting precedents for building standards today. The Environmental Protection Agency (EPA) highlights milestones in green building history, which include the following:

- American Institute of Architects (AIA) forms the Committee on the Environment Exit Disclaimer (1989)
- AIA publishes the EPA-funded *Environmental Resource Guide* (1992)
- EPA and the U.S. Department of Energy launch the ENERGY STAR program (1992)
- First local green building program is introduced in Austin, Texas (1992)
- U.S. Green Building Council (USGBC) Exit Disclaimer is founded (1993)
- "Greening of the White House" initiative is launched (1993)
- USGBC launches their Leadership in Energy and Environmental Design (LEED) Exit Disclaimer version 1.0 pilot program (1998)

Among the federal milestones are the Energy Policy Act of 2005, featuring federal building sustainable performance standards, and the Energy Independence and Security Act of 2007, which includes requirements for high-performance green federal buildings.

THE JOB

The EPA defines green building—also known as sustainable or high-performance building—as "the practice of creating structures and using processes that are environmentally responsible and resource-efficient throughout a building's life cycle from siting to design, construction, operation, maintenance, renovation, and deconstruction. This practice expands and complements the classical building design concerns of economy, utility, durability, and comfort."

Green builders aim to create energy-efficient (and low-maintenance) buildings at reasonable costs while conserving resources. They build healthy structures that have as little impact as possible on the environment. Commercial and residential buildings that are green are now identified through the LEED (Leadership in Energy and Environmental Design) designation. LEED is a universal rating system that the U.S. Green Building Council created in 2001 for sustainable standards through the life cycle of a building, including design, construction, operations, and maintenance.

Green builders may also be known as *construction or project managers, constructors, project engineers,* or *general contractors.* They coordinate, plan, and direct a variety of construction projects, such as the building of residential, commercial, and industrial structures, roads, bridges, wastewater treatment plants, and schools and hospitals. They may be responsible for the entire project or just a part of it. It's up to them to pull together a team of specialty trade contractors and oversee their work throughout the project. Managers might either own their own building company or work on staff with a company. Property owners, developers, and contracting firms overseeing the projects also contract construction managers.

Construction managers work with owners, engineers, architects, designers, and others, making sure that the planning, scheduling, and implementation of the construction design meets specifications. Complicated projects, such as industrial complexes and office buildings, entail larger teams and greater division of labor. For these projects, the specific types of work people are hired to do include the following:

- Site preparation, which often includes clearing land
- Creating or replacing sewage systems
- Landscaping and road construction
- Building construction, which includes excavation, foundation laying, and structural framework erection

Home builder Norm Schreifels stands outside his green home that he is building in Corrales, New Mexico. Schreifels and other contractors are using better building techniques and environmentally friendly materials to create what industry insiders are describing as the future of homebuilding. *AP Photo/Jake Schoellkopf*

❧   Building systems, such as fire protection, electrical, plumbing, air-conditioning, and heating

Green construction managers are also responsible for making sure that the materials used are green materials. They recycle and reuse building materials to reduce waste and conserve energy. Types of materials they use may be locally or regionally available; salvaged, refurbished, or remanufactured; and those having independent certification, such as certified wood. Construction managers figure out the best way to get the materials to the building site, keeping the project's budget and deadline in mind. They oversee worker productivity and safety, use of tools and equipment, and construction quality. They create the steps involved in the project and delegate and oversee the jobs. Depending on their role, they may select general and trade contractors for such work as plumbing, metalworking, painting, and carpet installation. It's also up to them to secure the building permits and licenses, and, in some cases, to make sure all construction activity meets insurance requirements.

REQUIREMENTS
High School
Take course work in math, science, and business. English and communications classes are also helpful because the work involves regular communication with clients, construction crews, architects, and a host of other team members. Art, design, and computer classes also provide a solid foundation in this type of work.

Postsecondary Training
Companies prefer to hire green builders who have a bachelor's degree in construction science, construction management, or civil engineering. Students can choose from 105 colleges and universities that offer bachelor's degree programs in construction science, building science, and construction engineering. Subjects covered include project control and development, site planning, design, construction methods and materials, value analysis, safety, building codes and standards, inspection procedures, engineering and architectural sciences, information technology, and more. Construction managers usually have a master's degree in construction management or construction science. Those who teach at colleges or conduct research usually have doctoral degrees.

Certification or Licensing

Certification is not required but it can enhance construction professionals' careers; clients view it as verification of competence and experience. Many green builders are certified as a LEED-Accredited Professional (LEED AP) through the Green Building Certification Institute. More than 75,000 construction professionals hold the LEED-AP designation.

Builders who meet educational and work requirements, and pass exams, can also receive professional constructors' certification through the American Institute of Constructors and the Construction Management Association of America.

Other Requirements

Green builders need a variety of skills to successfully handle their jobs. Clear verbal and written communication skills are essential in working with diverse groups of people. Fluency in Spanish has grown in importance because it is the first language of a growing number of construction workers. Projects can change midway; construction problems can arise; budget issues may occur—issues like these require flexibility and decisiveness. The ability to manage multiple projects at once, meet deadlines, work within budgets, and effectively lead teams are key components of a green builder's job.

EXPLORING

Read as much as you can about green building to keep up with latest developments and trends. Search the Internet for books, magazines, and videos about sustainable building. And visit the Web sites of green building companies to get a better sense of the types of projects they work on and who their clients are. Two companies to start you off in your research are New York City-based Green Street (http://www.greenstreetinc.com) and Maryland-based Green Builders Inc. (http://www.greenbuilders.com).

EMPLOYERS

About 487,000 construction managers were employed in the United States in 2006. More than half were self-employed as construction firm owners. Most of those that were on staff worked in the construction industry, with 13 percent working for specialty trade contractor businesses (e.g., plumbing, heating, air-conditioning, and electrical); 9 percent in residential building construction; and 9 percent in nonresidential building construction firms. Other construction managers worked for the government and for architectural, engineering, and related services firms.

STARTING OUT

Working on a green building project is the way to test the waters while gaining valuable skills. Search the career section of the U.S. Green Building Council's Web site (http://www.usgbc.org) to learn more about job and intern opportunities.

ADVANCEMENT

Green builders who work for construction companies can advance by assuming more responsibilities, handling more complex projects, and moving up the ranks to management or executive positions. If they already hold a master's degree in construction management, they can enhance their marketability through a graduate degree in business administration or finance. Builders can also advance by becoming independent consultants and starting their own building companies.

EARNINGS

Green builders' salaries vary according to work experience and job title. In 2008 the median income of construction managers was $79,860. The bottom 10 percent earned $47,000 or less, and the highest 10 percent earned $145,920 or more. Those who worked in nonresidential construction earned higher salaries, with a median income of $88,550. Foundation, structure, and building exterior contractors averaged $90,260. Cost estimators had salaries ranging from $33,150 to $94,470. California, Massachusetts, New York, New Jersey, and Washington are among the top-paying states for construction professionals.

WORK ENVIRONMENT

Green builders work in offices and at construction sites. For overseas projects, they may need to temporarily relocate to another country. Builders work long hours, often more than 40 hours per week. They are on call 24 hours a day, seven days a week, dealing with emergencies, work delays, and issues due to bad weather. When they are at construction sites, they need to be careful and constantly aware of their surroundings.

OUTLOOK

Green builders should have excellent job opportunities in the years to come. Green building is on the rise due to concerns about the environment and human health. Sustainable building practices are becoming the norm, and as a result, green build-ers are in high demand. The U.S. Department of Labor predicts that the employment of construction managers, including green builders, will be faster than the average for all other occupations through 2016.

Construction projects are becoming more complex due to laws regarding standards for buildings and construction materi-als, worker safety, energy efficiency, environmental protection, and construction litigation. Building professionals with experi-ence and knowledge of these areas will be in particular demand. Also, those who are skilled in using advanced technology will find more employment opportunities. More multipurpose build-ings and energy-efficient structures are being commissioned, and portions of U.S. infrastructure are continuously being replaced, creating job opportunities for building professionals. Construc-tion managers will be needed to help build residential homes, office buildings, shopping malls, hospitals, schools, restaurants, and other structures.

FOR MORE INFORMATION

For information about certification as a construction professional, visit the Web sites of the two following organizations:

American Institute of Constructors
717 Princess Street
Alexandria, VA 22314-2221
Tel: 703-683-4999
http://www.aicnet.org

Construction Management Association of America
7918 Jones Branch Drive, Suite 540
McLean, VA 22102-3366
Tel: 703-356-2622
http://www.cmaanet.org

To learn more about environmental issues, the latest trends, and volunteer opportunities, visit the EPA's Web site.

Environmental Protection Agency (EPA)
Ariel Rios Building
1200 Pennsylvania Avenue, NW
Washington, DC 20460-0001
Tel: 202-272-0167
http://www.epa.gov

For information about LEED professional and project certification, contact

Green Building Certification Institute
2101 L Street, NW, Suite 650
Washington, DC 20037-1526
Tel: 202-828-1145
http://www.gbci.org

To learn more about green buildings, contact

Sustainable Buildings Industry Council
1112 16th Street, NW, Suite 240
Washington, DC 20036-4818
Tel: 202-628-7400
Email: sbic@sbicouncil.org
http://www.sbicouncil.org

For information about workshops, publications, and membership, contact

U.S. Green Building Council
2101 L Street, NW, Suite 500
Washington, DC 20037-1526
Tel: 800-795-1747
Email: info@usgbc.org
http://www.usgbc.org

Green Products Manufacturers

QUICK FACTS	
School Subjects Business Communications English	**Minimum Education Level** Bachelor's degree
	Salary Range $50,330 to $83,290 to $140,530+
Personal Skills Analytical Creative Decisive	**Certification or Licensing** Voluntary certification is beneficial
Work Environment Primarily indoors Primarily one location	**Outlook** About as fast as the average

OVERVIEW

Green products manufacturers make products that are environmentally friendly. They may create products made from recycled materials or products that use water and energy efficiently.

HISTORY

Prior to the 1700s, skilled artisans, with the help of assistants, manufactured products. They used their hands and tools to make such items as candles, carriages and wagons, dresses, and pots. They were also ironworkers, bakers, or tobacconists. They belonged to craft guilds and learned their trade through apprenticeships.

From the mid-1700s to the late 1800s, the first industrial revolution occurred in northern Europe and quickly spread to the rest of the world. During this time, the development of the steam engine, the sewing machine, Eli Whitney's cotton gin (a machine that separated cotton seeds from fiber), and other innovations changed manufacturing processes and improved productivity. Following the Civil War, in the 1880s, new inventions and discoveries abounded in the United States. This second industrial revolution gave rise to railroad expansion, telephones, phonographs, typewriters, electric lights, and cars. The industrial infrastructure had early roots at this time as well, with the discovery of coal and oil. Large iron, steel, copper, and silver mines were opened, closely followed by lead mines and cement factories. Mass-production methods were developed and by 1910, electricity had replaced waterpower, thus speeding up the production process.

Henry Ford, founder of the Ford Motor Company, was an early pioneer of large-scale manufacturing in the early twentieth century. Ford created assembly lines to improve workflow and productivity and realized the importance of workers and paid them high wages. Ford focused not just on the product but also on the production process.

World War II triggered another spike in the manufacturing industry. Soldiers at war needed equipment and machinery, and manufacturers had to develop products and manufacturing methods to produce and deliver what was needed quickly and efficiently while still maintaining quality control. Until the 1940s, manufacturing was predominantly based in the Northeast and the Midwest. Since World War II, manufacturing companies have opened in the West and South.

The U.S. manufacturing industry has had its ups and downs over the past 50 years, with many companies moving operations overseas to save money. But some believe manufacturing is still alive and evolving in this country. Writer Stephen Manning, in his article "U.S. Manufacturing Goes Upscale" (Associated Press, February 16, 2009), stated that America "by far remains the world's leading manufacturer by value of goods produced. It hit a record $1.6 trillion in 2007—nearly double the $811 billion in 1987. For every $1 of value produced in China's factories, America generates $2.50."

THE JOB

More people are seeking out green products because of heightened awareness of environmental issues such as greenhouse gas emissions, global warming, overflowing landfills, and pollution. As a result, more manufacturers are turning their businesses into sustainable businesses. Green products manufacturers are concerned about the impact that manufacturing production processes, and use of the products themselves, have on the environment. To tackle this problem, many are transforming recyclable materials into products that won't negatively affect the earth. They are also creating products that use water and energy more efficiently, such as low-flow toilets, solar panels, and hybrid cars.

Manufacturing is the process of turning raw materials into usable products. Depending on the scale and scope of the business, manufacturing can be done by hand or by machine. The main areas manufacturing covers include food; textiles, clothing, and footwear; metals; printing and paper; furniture and cabinet making; machinery and equipment; and petroleum, coal, chemical, and related products. Green products can be made from recycled or reused materials such as glass bottles and jars, plastic bottles and milk jugs, yogurt containers, aluminum and tin cans, bottle caps, newspapers and magazines, cardboard boxes and containers, blue jeans, and even old car tires.

One example of a manufacturing company concerned about the environment is Homasote, a New Jersey-based manufacturer of building products made from recycled materials. According to its Web site (http://www.homasote.com), the company's manufacturing process recycles up to 250 tons of postconsumer paper (including newspapers), helps conserve nearly 1.5 million trees, and eliminates more than 65 million pounds of waste that would otherwise be in landfills. Homasote also recycles all of the water it uses to manufacture its products.

The Subaru of Indiana (SIA) plant (http://www.subaru-sia.com), located in Lafayette, is another example of a manufacturing company that is actively working to reduce its imprint on the environment. It recycles and reuses 97 percent of its refuse, and it boxes and ships the remaining 3 percent to the city of Indianapolis, where the refuse is incinerated to help generate steam. As a result, nothing is sent to landfills, which is why SIA achieved a "zero landfill" status. To help people understand how dramatic this business practice is, SIA's Web site states: "When you carry out the trash on the next collection day, you're sending more to landfills than does the SIA plant in Lafayette." Types of things SIA is recycling and/or reusing include: brass lug nuts (previously, 33,000 pounds of brass was thrown away each year); paint sludge, which is dried into powder, then shipped to plastics manufacturers to reuse for guardrails and parking-lot bumpers; and solvents, which are cleaned and reused in paints.

An HNI Corporation worker paints a piece of office furniture on the production line. Executives at HNI have found that recycling wood, cardboard, paint, and fabric saves money, boosts the company's "green credentials," and pleases customers. *AP Photo/Charlie Neibergall*

The types of jobs involved in manufacturing include *product designer, manufacturing production manager, assembler and fabricator, quality control inspector,* and *manufacturing sales representative.*

Product designers focus on style, function, quality, and safety when they design products. They usually specialize in one area, such as cars, appliances, housewares, toys, technology goods, or pharmaceuticals. They create designs by hand and with computer-aided design (CAD) and computer-aided industrial design (CAID) software. They make sure the design meets company specifications.

Manufacturing production managers oversee one area of the manufacturing process if the plant is a large operation; if it's a small business, they may be responsible for managing the entire plant. Managers plan, direct, and coordinate the production process. They create the plan for the work, deciding on the sequence of work, the staff and machines needed, and the production schedule. They work closely with other departments to make sure budgets are adhered to, deadlines and goals are met, and company policies are followed.

Assemblers and fabricators use tools, machines, and their hands to put products together. They assemble automobiles, airplanes, household appliances, computers, electronic devices, and more. Their work may be easy, or it may be complex, requiring them to closely read and follow schematics and blueprints.

Quality control inspectors check products to make sure they meet specific standards. The nature of their work depends on the types of products they inspect. For example, materials inspectors will check products by sight, smell, sound, feel, and even taste. They look for imperfections such as scratches, cuts, bubbles, missing parts, and stitching and seam errors. They may also check measurements and make sure products operate correctly.

Sales representatives promote products to wholesale and retail buyers and purchasing agents. They demonstrate their products to prospective clients, explaining how they can reduce costs and drive up sales, and address questions and concerns.

Other jobs in manufacturing include *cost estimators*, who compile data on all the factors that can impact costs involved in manufacturing a product; *production clerks*, who help with the flow of information, work, and materials among manufacturing offices; and *machinists*, who use machines to create parts for products.

REQUIREMENTS
High School
Manufacturers bring a variety of skills to their jobs. Course work in business, math, English, science, art, and computers will provide a well-rounded foundation. Shop classes will benefit individuals interested in pursuing work as machinists or other related jobs.

Postsecondary Training
Many manufacturers have an undergraduate degree. Depending on the job, individuals may have a bachelor's degree in business administration, management, industrial technology, industrial design, or industrial engineering. Some companies may hire individuals with other types of undergraduate degrees providing their skills, work experience, and personality match the job requirements. Individuals with graduate degrees in business administration or industrial engineering will find more opportunities with larger manufacturing companies.

Certification or Licensing
Certification is not required but can give a manufacturing employee a boost in the job market. The Association for Operations and Man-

agement offers the CPIM (Certified in Production and Inventory Management) designation to individuals who pass a series of tests and who continue to complete specific work activities every three years. The American Society for Quality offers the Certified Manager of Quality/Organizational Excellence (CMQ/OE) credential to managers who have at least 10 years of experience or relevant education and who pass an exam.

Other Requirements

Depending on the job, manufacturing employees need solid communication and math skills, good business and sales sense, analytical and decision-making abilities, creativity, design software knowledge, good hand-eye coordination, and self-discipline. Knowledge of and appreciation for sustainable manufacturing processes is also essential.

EXPLORING

Learn more about manufacturing by watching educational videos on Alabama Public Television's Web site (http://www.aptv.org). Attending a green manufacturing expo is also an excellent way to see all the different types of green products and manufacturing companies under one roof. For listings of upcoming events, visit Green Manufacturing Expo's Web site (http://www.devicelink. com/expo/gmx10).

EMPLOYERS

In 2006 there were 157,000 production managers employed in the United States, according to the U.S. Department of Labor. About 80 percent of them worked for manufacturing industries. Assemblers and fabricators held about 2.1 million jobs. Quality inspectors, testers, weighers, and samplers held about 491,000 jobs. Commercial and industrial designers held about 48,000 jobs. More than 50 percent worked for manufacturing companies, 30 percent were self-employed, and 15 percent worked for engineering or design firms.

STARTING OUT

One way to get a foot in the door at a manufacturing company is through an internship or apprenticeship. Search the Internet for green manufacturing companies near you and contact them to see if they need any part-time office help. Some manufacturing employees get started through on-the-job training. Find out if there are any openings for summer help and if they offer training programs.

ADVANCEMENT

Manufacturers who own their own plants can expand by developing and producing new products and growing their business. Those who work for manufacturing companies can advance by taking on more responsibilities, managing larger staffs, and moving up to supervisor and management positions. Another way to advance is to learn new skills or sharpen current ones by taking continuing education classes, pursuing an advanced degree, or getting certified in a specific area.

EARNINGS

Salaries for manufacturing employees vary depending upon their job title and experience. In 2006 production managers had median annual incomes of $83,290, with the lowest 10 percent earning $50,330 and the highest 10 percent earning $140,530 or more. Industrial designers had salaries ranging from $31,400 to $97,770. Quality inspectors had median incomes of $31,240. Salaries for assemblers and fabricators varied according to the industry in which they worked; for example, in 2006, aircraft assemblers and fabricators had median incomes of $44,130 and $32,660 per year, respectively; and engine and machine assemblers and electrical and electronic assemblers averaged $27,490 per year. Sales representatives for wholesale and manufacturing companies (not including technical and scientific products) had median annual salaries of $51,330, with the bottom 10 percent earning $26,950 and the top 10 percent averaging $106,040 or more.

WORK ENVIRONMENT

Products manufacturers work in offices and plants. They may work 40-hour workweeks during business hours, or they may work nights and weekends depending on their work shift. Some manufacturing employees travel for client meetings and plant inspections. Some may also travel for conferences, expos, workshops, and training programs.

OUTLOOK

General manufacturing is expected to show a decline in employment through 2016, according to the U.S. Department of Labor. Green products manufacturers should have better employment opportunities during the coming years, however, due to increased consumer demand for green products. In addition, local, state, and federal laws are requiring companies to reduce carbon emissions, thus causing more plants to "green" their operations. Individuals with green production manufacturing knowledge and experience will be in greater demand.

FOR MORE INFORMATION

For information about manufacturing in the United States, contact

Alliance for American Manufacturing
27 Fifteenth Street, NW, Suite 700
Washington, DC 20005-2168
Tel: 202-393-3430
Email: info@aamfg.org
http://www.americanmanufacturing.org

For information about quality inspectors, visit

American Society for Quality
600 North Plankinton Avenue
Milwaukee, WI 53203-2914
Tel: 800-248-1946
http://www.asq.org

For information on production management and certification, contact

APICS Association for Operations Management
8430 West Bryn Mawr Avenue, Suite 1000
Chicago, IL 60631-3417
Tel: 800-444-2742
http://www.apics.org

For information about the recycling process and recycled products, look for articles on EDF's Web site.

Environmental Defense Fund
Membership and Public Information
1875 Connecticut Avenue, NW, Suite 600
Washington, DC 20009-5739

Tel: 800-684-3322
http://www.edf.org

To learn more about commercial and industrial design, contact

Industrial Designers Society of America
45195 Business Court, Suite 250
Dulles, VA 20166-6717
Tel: 703-707-6000
http://www.idsa.org

Green Recruiters

OVERVIEW

Green recruiters are employment specialists who place people in environmental jobs. They help people find positions in administrative and office support, communications, public relations, marketing, development, science, technology, and human resources. They may also provide consulting services to organizations seeking recruitment advice.

HISTORY

In the 1700s and 1800s many workers belonged to crafts guilds and formed trade unions as ways to improve working standards and voice their rights. It wasn't until the early 1900s that job design principles and work practices were examined more closely.

Frederick Taylor's book *The Principles of Scientific Management*, published in 1911, introduced the concept of using scientific management in the workplace. Taylor was an advocate of work standardization, work measurement, and production and quality control. By following these principles, workers are assessed and chosen scientifically; managers must cooperate with workers to ensure the job is done based on the way it was taught; and managers and workers have distinctly different, clearly defined job responsibilities. These principles remove worker initiative, creativity, and individuality, yet increase production efficiency and ensure that production technology operates correctly. American industries applied these principles throughout the twentieth century and prospered enormously.

By the late 1920s, as a way to increase worker motivation and productivity, more attention was being paid to treating workers as human beings as opposed to machines. In the Hawthorne Studies, which were experiments that Australian-born psychologist Elton Mayo conducted (1927–1932) at the Western Electric Hawthorne plant in Chicago, small groups of workers were studied to discern the relationship between productivity and the work environment. Mayo found that the work environment had a direct impact on productivity, and that workers who felt supported by their managers performed better than those who experienced low morale. The human relations movement of the late 1920s and 1930s stemmed from these studies. The movement came about when a group of social scientists studied the behaviors of people in workplace groups, focusing on individuals and their emotions. The need for increased communications between workers and managers has since become a key element in productive and healthy work environments.

These early studies and findings helped form the basis for today's human resources industry. Recruiting and selecting appropriate applicants, motivating employees and reviewing job performance, monitoring workers, improving industrial relations, offering employee benefits packages, and providing career advancement opportunities through training and development—all of these things have evolved because people realized that productive, happy employees contribute to a better environment overall.

THE JOB

Green recruiters help match socially responsible companies with employees who share their values. Recruiters may specialize in environmental sciences, sustainable development and corporate change, and corporate social responsibility. Successful recruitment and placement requires a deep appreciation of the industry, the skills needed to successfully handle the job, and the personality and character traits that mesh best with the company's philosophy and environment. As a result, most recruiters have educational and work backgrounds in the areas for which they recruit.

Job responsibilities for green recruiters will vary based on the size of the company they work for. Those at small agencies may be *human resources generalists*, handling a variety of human resources tasks, which could be everything from writing and placing recruitment advertisements to reviewing resumes and interviewing candidates. Those who work for larger firms may report to a *human resources director*, who oversees several departments that are usually headed by *human resources managers*. Managers generally specialize in employment and placement, compensation and benefits, training and development, or labor relations. They supervise the hiring of employees, and they also oversee employment, recruitment, and placement specialists.

Green recruiters actively search for the best candidates for jobs. They travel often to meet with business owners and develop and maintain relationships within the green business community. Recruiters also travel to college campuses and job fairs, always looking for the most qualified job applicants. Their responsibilities may entail screening, interviewing, and testing applicants; checking references; negotiating salaries; and extending job offers. The types of things they discuss with job applicants include the company's history and mission, the job description, terms of the job, the work environment and company policies, salary, and promotion opportunities.

Many employment firms offer placement services in multiple sectors and are adding environmental placement to help meet growing demand in this arena. Maryland-based employment firm Aerotek, founded in 1983, is such a firm. It recruits and places for automotive, aviation, energy, scientific, and environmental jobs, to name a few. The types of environmental jobs Aerotek helps recruit for include air quality specialists, archaeologists, soil scientists, field chemists, general laborers, heavy equipment operators, inspection personnel, and site engineers.

Bright Green Talent is an example of a recruitment firm focused solely on helping environmental companies find like-minded employees. The mother company Bright Green was founded in London in 1999 by Tom Hannam, a Don (Fellow) at Oxford, and Tom Savage, a social entrepreneur who founded Blue Ventures, an award-winning marine conservation organization. They each wanted to create a business that helped the environment and society by offering rewarding career opportunities to socially concerned people. They place people in jobs ranging from engineers and scientists to sales managers, graphic designers, and even interns. They also offer a blog, webinars, and career coaching.

REQUIREMENTS

High School

Recruiters need to have excellent communication skills to fully understand what companies are seeking in employees, and to clearly explain the details to job applicants. Having a background in the areas they're recruiting for is also recommended. Course work in business, English, and the sciences will provide a solid foundation. Computer classes are also extremely useful.

Postsecondary Training

Educational backgrounds vary in this field because recruiters work in a range of industries. Some green recruiters have a bachelor's degree in human resources, human resources administration, or labor relations. Many others have degrees in the specific areas they recruit and place for, such as social or behavioral science, business, engineering, science, finance, or law. For a well-rounded program of study, be sure to also take classes in compensation, recruitment, training and development, performance appraisal, principles of management, organizational structure, and industrial psychology. Classes in labor law, business administration, psychology, sociology, political science, economics, and statistics are also beneficial.

For management positions within larger companies, most recruiters have a master's degree in human resources, labor relations, or business administration. A Ph.D. is helpful for those who teach, write, and consult.

Certification or Licensing

Certification is not required to do this job, but it can enhance a recruiters' career. Organizations such as the Society for Human Resource Management, the American Society for Training and Development, the International Foundation of Employee Benefits Plans, and the World at Work Society of Certified Professionals offer certification to individuals who meet specific educational and work experience requirements and pass certification exams. Designations include certified employee benefits specialist, professional in human resources, and work-life certified professional.

Other Requirements

Recruiters need to have many diverse skills to succeed in this field. In addition to having an educational and work background in the specific area in which they work, they need to be able to communicate clearly and effectively. Strong written and verbal communication skills are essential. They email, phone, and meet regularly with clients, job applicants, and their own staff members, so they must enjoy working with people. Diplomacy, honesty, integrity, and discretion are also important in this type of work.

EXPLORING

There are a myriad of green recruitment agencies and the numbers are growing every day. You can learn more about the types of industries and jobs they recruit for by for visiting their Web sites. Use a search engine such as Google and plug in search words like "green recruitment firms" and "environmental recruitment." You can also keep up with business trends and employment growth forecasts by reading the *New York Times*, the *Wall Street Journal*, *Forbes*, and *BusinessWeek*.

EMPLOYERS

About 868,000 human resources, training, and labor relations managers were employed in the United States in 2006. Of those, 197,000 were employment, recruitment, and placement specialists; and 136,000 were human resources managers. About 17,000 managers and specialists were self-employed.

Recruiters and managers work in all industries. About 90 percent work in the private sector, which includes administrative and support services; professional, scientific, and technical services; health care and social assistance; finance and insurance firms; and manufacturing. About 13 percent of recruiters work as human resources managers and specialists for the government. They recruit, interview, classify jobs, train, administer salaries and benefits, manage employee relations, and handle other employment-related tasks for public employees.

STARTING OUT

Working in a human resources department is a great way to learn more about job recruitment and placement. Make a list of companies that interest you and contact their human resources departments to see if they offer internship programs or need part-time office help.

ADVANCEMENT

Recruiters can advance from human resources generalists to positions of higher authority, such as managers and directors. They may oversee more departments and larger staff, and help create new recruitment departments in other branch offices. They may also start their own recruitment agencies. Those that own their own businesses can advance by growing their businesses, adding more staff members, expanding into other sectors and specialty areas, and by consulting, writing, and teaching.

EARNINGS

Salaries for recruiters vary depending on their position. According to the U.S. Department of Labor, in 2008 employment, recruitment, and placement specialists earned median annual salaries of $45,470, with the lowest 10 percent earning $28,030 and the highest 10 percent earning $85,760. Human resource managers earned salaries ranging from $56,770 to $163,440 in 2008.

Recruiters who work on staff with agencies may have other benefits as well, including salary raises, bonuses, health insurance coverage, paid vacation and sick time, and professional education reimbursement.

WORK ENVIRONMENT

Recruiters work in offices or in their homes. They travel for client meetings, and to college campuses and job fairs. Most work 35- to 40-hour workweeks, with work extending to evenings and/or weekends during campus recruitment and employment conference trips.

OUTLOOK

Human resources specialists and managers are expected to have faster than average employment opportunities through 2016, according to the U.S. Department of Labor. Green businesses are growing. The drive for clean energy is creating a need for more employees in this sector. In fact, many states are introducing clean energy initiatives that are spurring job growth. And more companies are hiring staff to help them address corporate social responsibility issues.

FOR MORE INFORMATION

To learn more about green jobs and recruitment, visit the Web sites of these recruitment firms.

Aerotek Inc.
7301 Parkway Drive
Hanover, MD 21076-1159
Tel: 410-694-5100
http://www.aerotek.com

Bright Green Talent
200 Pine Street, Suite 400
San Francisco, CA 94104
Tel: 415-391-0729
Email: speakout@brightgreentalent.com
http://brightgreentalent.com

To learn more about workshops and certification programs in human resources, contact these organizations.

American Society for Training and Development
1640 King Street, Box 1443
Alexandria, VA 22313-2043
Tel: 703-683-8100
http://www.astd.org

International Foundation of Employee Benefit Plans
18700 West Bluemound Road
Brookfield, WI 53045-2936
http://www.ifebp.org

Society for Human Resource Management
1800 Duke Street
Alexandria, VA 22314-3494
Tel: 802-283-SHRM
http://www.shrm.org

World at Work
14040 North Northsight Boulevard
Scottsdale, AZ 85260
Tel: 480-922-2020
Email: customerrelations@worldatwork.org
http://www.worldatwork.org

Grounds Maintenance Workers

QUICK FACTS

School Subjects
Art
Business
Horticulture

Personal Skills
Artistic/creative
Mechanical ability
Physical fitness

Work Environment
Primarily outdoors
One or multiple locations

Minimum Education Level
High school education

Salary Range
$16,680 to $22,390 to
$47,830

Certification or Licensing
Most states require licensure
for pesticide usage

Outlook
Faster than the average

OVERVIEW

Grounds maintenance workers maintain lawns, gardens, and grounds. They may care for indoor and outdoor gardens and plantings at business and apartment buildings, schools, museums, libraries, cemeteries, and even malls and hotels. They use a variety of hand tools and equipment in their work, including power lawnmowers, chainsaws, electric clippers, snow blowers, and sometimes tractors and twin-axle vehicles. They work long days outside in most weather conditions, and must be physically fit to meet all the demands of the work.

HISTORY

Until the end of the 18th century "lawn care" was not a term in America's vocabulary. Lawns back then were usually dirt-packed, and the gardens that existed were small and compact, and not for show. These "cottage gardens," as they were dubbed, were productive and practical, generally filled with flowers, herbs, and vegetables, which could be used for food, medicine, and dyes. It was the British and the Scots, in the late 1700s, who inspired wealthy Americans to "green" their yards and adorn their lawns. After visiting England and Scotland and seeing the beautiful, sprawling landscapes of the estates (and the healthy grasses of the golf and bowling lawns), Americans returned to the United States and made it their mission to recreate these images.

From the 1800s to early 1900s, wealthy Americans, institutions, and the U.S. government were the main employers of grounds maintenance workers, retaining them to tend to lawns, grounds, and gardens. Mechanical mowing came about in the mid-1800s, and in 1870, Elwood McGuire of Richmond, Indiana, created the push mower. By the end of the 19th century, the United States was building 50,000 lawnmowers per year and shipping them to countries around the world.

Outdoor recreational sports started drawing great interest in the early 1900s, creating need for grounds maintenance workers to help install and maintain playing fields. Interest in lawn care intensified when, from 1910 to 1924, the U.S. Department of Agriculture, in collaboration with the U.S. Golf Association, experimented with different types of grasses in an effort to create grass that could endure different climates throughout America. In fact, the Pentagon sits on the spot where the first experimental U.S. turf farm was reportedly located.

By the mid-1900s, cities were growing and parks were being built as havens for the burgeoning population. Landscaping design and materials grew in importance, as did the need for landscaping designers and maintenance workers. For the past 50 years, a lawn of one's own has been, and continues to be, the American dream. In 2000, according to a Gallup survey, 26 million American households hired a landscaping professional, and that number continues to grow.

THE JOB

Grounds maintenance workers, also known as *landscapers*, *groundskeepers*, or *gardeners*, maintain lawns, gardens, landscapes, and plants for a variety of clients. Their tasks include mowing, watering, fertilizing, pruning, and weeding. They install flowers, plants, lighting, and sprinkler systems based on landscape design specifications. They may work on a small crew with minimal supervision or on a large team with several managers overseeing their work.

Grounds maintenance workers use equipment such as power saws, mowers, tractors, aerators, weed whackers, handsaws, clippers, and more; many also know how to repair this equipment. Some are trained tree climbers and pruners; you'll often see them alongside roads, high up in cherry pickers, clipping dead tree branches or limbs that are leaning on or close to telephone and power lines. Others are trained and licensed in pesticide application.

Day-to-day responsibilities will vary depending on where grounds maintenance workers work. Those who tend sports fields, golf courses, parks, or other landscapes that feature turf will work under the direction of *turf managers*. Tasks involved include aerating the soil, reseeding and fertilizing existing turf, as well as mowing, watering, and weeding the grass. Turf managers are also responsible for making sure the appropriate marks are drawn on the turf, in accordance with the sport, and that equipment, such as goalposts for games, is set up. Workers help install, program, and repair irrigation systems to ensure that the turf stays healthy and strong.

Landscape contractors and *supervisors* run their own businesses. It's up to them to interview and select workers; negotiate salaries and wages; purchase and maintain equipment and supplies (e.g., plant materials); establish work schedules; and supervise workers. In addition to all of this, they have to attract and maintain relationships with clients; they promote their services through their Web site, print advertisements, and other media. They meet with clients and landscape designers to discuss plans and review drawings. They then create cost estimates that fall within the client's budget. Landscape contractors may also handle the bookkeeping themselves or have full- or part-time office staff tend to the administrative tasks.

Some landscape contractors may also be trained designers—this gives them an edge in the market because they can bring design and building knowledge to projects. Using computer-aided design software, they create landscape designs, adding in such things as trees, plants, shrubs, walkways, and walls. They either install these elements themselves or oversee the installation.

REQUIREMENTS
High School
While there are no educational requirements for this job, high school course work focusing on art, design, science (ecology and geology), and English will provide a well-rounded foundation. Classes in business and math will also be useful for those wishing to move up in this field.

Postsecondary Training

Most grounds maintenance workers learn while on the job, usually during the first few weeks. They learn such things as how to drive lawnmowers or small tractors, how to operate leaf blowers, and what safety precautions to take. Some large businesses, such as golf courses or municipalities, may offer on-the-job training as well as cover the expense of worker education in areas such as horticulture or small-engine repair.

Small business owners and specialists in grounds maintenance usually have bachelor's degrees, learning more specifically about landscape design, horticulture, arboriculture, and business management.

Certification or Licensing

Most states require that grounds maintenance workers who use insecticides, herbicides, and fungicides be licensed or certified. They must pass a test showing they understand how to properly use and dispose of these materials. Some states require that landscape contractors be licensed.

While not required, certification in areas aside from pesticide knowledge can offer career advancement opportunities. The Professional Grounds Management Society (PGMS) provides voluntary certification to grounds managers who have a bachelor's degree, at least four years of work experience, and who have passed the certification exam. The Professional Landcare Network (PLANET) also offers certification in such areas as landscaping and grounds maintenance. And the Tree Care Industry offers such credentials as ground operations specialist and even tree climber specialist. Many professional organizations offer certification programs in safety as well.

Other Requirements

Mowing lawns, hedging bushes, and climbing trees, day in and day out, can take a toll on the body. Grounds maintenance workers need to be physically fit, and must take care of themselves so that they stay healthy and able to work. Being responsible and motivated is important in this job, and having good communication skills will help in dealing with customers as well as with fellow workers and crew supervisors.

EXPLORING

The best way to learn about this job is firsthand. Join a garden club so you can start learning more about plants. If you have a lawn, start helping your parents to maintain it, or ask your neighbors if they need help taking care of theirs. You can start by doing the basics: mowing, weeding, hedge trimming, and gardening. Talk to landscaping business owners near you and find out if they need any part-time help. If you feel ambitious, you may even want to commit the summer to working full-time hours, just to see if this is something you'd be interested in doing for the longer term. (It will be a way to gauge your stamina and physical fitness, as well.) You can also learn more about this type of work by visiting the Professional Grounds Management Society's Web site (http://www.pgms.org).

EMPLOYERS

According to the U.S. Department of Labor, approximately 1.5 million grounds maintenance workers were employed in the United States in 2006, with the majority working in landscaping and groundskeeping. Companies that offer landscaping services to buildings and dwellings employ approximately one-third of all grounds maintenance workers. Many work for parks, hospitals, golf courses, racetracks, educational institutions, and local governments.

One in four grounds maintenance workers is self-employed, working on a contract basis. About 14 percent work part time, and about 9 percent are under 20 years of age.

STARTING OUT

Most grounds maintenance workers get their start in this field by taking a part-time or full-time job with a lawn care or landscaping service. Others start out in high school with their own lawn-mowing business, working in their neighborhood and then branching out through word of mouth.

ADVANCEMENT

Grounds maintenance workers who own their own business usually advance by expanding their client base, taking on new and different projects, and/or becoming specialists in other lawn care services. Those who work for companies who have strong communication skills and demonstrate reliability and excellent work can advance to supervisory positions such as crew leader. Furthering your education in the field through certification programs, such as those with PGMS and PLANET, is another way to advance. "Master gardener" training programs, offered by agricultural cooperative extension services of state land-grant universities, are useful to those who want to further their careers by learning more about botany, composting, organic gardening, pesticide use and safety, and soils and fertilizers.

EARNINGS

In 2008 landscaping and groundskeeping workers earned median annual salaries of $23,150, with the lowest 10 percent earning $16,600 or less annually, and the highest 10 percent earning $36,550 or more. Pesticide handlers, sprayers, and applicators earned slightly higher salaries: the median annual income was $29,770; the lowest 10 percent earned $19,820, and the highest 10 percent earned $44,910. Tree trimmers and pruners had similar earnings: the lowest 10 percent earned $20,000; the median income was $29,970; and the highest 10 percent earned $46,680. All other grounds maintenance workers had annual incomes ranging from $16,680 to $47,830.

WORK ENVIRONMENT

Grounds maintenance workers work primarily outdoors, unless the plantings and gardens are located inside buildings and covered areas. They may work at people's homes and residential development properties, commercial buildings, industrial parks, hospitals, amusement and recreation facilities, or the green areas alongside highways. They usually work during the spring, summer, and fall. Those who work in more temperate environments work all year round.

OUTLOOK

The future looks very rosy for grounds maintenance workers. The U.S. Department of Labor predicts faster than average growth in employment opportunities through 2016, with roughly 270,000 new jobs opening up in that timeframe. Lawn care and landscaping companies will need help to continue meeting current customers' lawn care needs, as well as to meet growing demand due to increased construction of office buildings, shopping malls, residential housing, highways, and parks. Utility companies will also need grounds maintenance workers to trim and prune trees that interfere with power lines.

More people are paying attention to their front lawns and backyards, wanting to beautify them in ways that work with the environment. The media, particularly television shows focusing on home makeovers and renovation and "green" living, has raised awareness and inspired people to hire lawn care services. In addition, because the work is so physically demanding and most starting salaries for beginners are typically low, companies are challenged in attracting and keeping employees. As a result, there are far more job openings than there are grounds maintenance workers to fill them.

FOR MORE INFORMATION

See if there's a garden club near you, and get gardening tips and ideas for projects by contacting

National Garden Clubs Inc.
4401 Magnolia Avenue
St. Louis, MO 63110-3406
Tel: 314-776-7574
Email: headquarters@gardenclub.org
http://www.gardenclub.org

For information on grounds management educational programs, events, and news, contact

Professional Grounds Management Society
720 Light Street
Baltimore, MD 21230-3850
Tel: 410-223-2861
Email: pgms@assnhqtrs.com
http://www.pgms.org

For business development advice, industry trends, publications, and networking opportunities, visit PLANET's Web site.

Professional Landcare Network (PLANET)
950 Herndon Parkway, Suite 450
Herndon, Virginia 20170-5528
Tel: 703-736-9666
http://www.landcarenetwork.org

Land Acquisition Professionals

QUICK FACTS

School Subjects
  Business
  Earth science

Personal Skills
  Communication/ideas
  Leadership/management

Work Environment
  Indoors and outdoors
  One location with some
    travel

Minimum Education Level
  Bachelor's degree

Salary Range
  $21,860 to $46,130 to
    $102,250+

Certification or Licensing
  None available

Outlook
  About as fast as the average

OVERVIEW

Land acquisition professionals assist in efforts of nonprofit land trusts to preserve land and water from development, subdivision, overly heavy recreational or agricultural use, or other human disruption by handling the land transaction—buying the land outright, acquiring development rights to it, obtaining easements, getting landowners to donate the land, or similar actions.

HISTORY

Land acquisitions has evolved as a specialty within nonprofit land trusts, which in turn are a special part of land and water conservation efforts in this country.

Land and water conservation efforts in the United States go back more than 100 years, when the federal government first started setting aside wilderness areas and other open land and bodies of water. Since then, hundreds of millions of acres have been preserved in federally owned and managed national parks, wildlife refuges, wild and scenic rivers, and other areas as well as state or locally managed protected lands. Today, acquisition by the federal government is largely complete, but acquisitions by private land trusts continue. The Land Trust Alliance reports that local and regional private, nonprofit land trusts have conserved more than 11.9 million acres of open space as of 2005—double the acreage protected as of 1998.

Broadly, land trusts are private nonprofit groups formed to acquire and manage open lands for the public's benefit. The first official one in this country was the Trustees of Reservations, formed in Boston in 1891. Concerned that open lands around the city were being rapidly swallowed up by development, this group of private citizens took action. They bought up some land themselves and made it available to the public for recreation. The Trustees of Reservations still exists today, and is still acquiring land statewide. In fact, Massachusetts has the largest number of land trusts of all the states. But land trusts also exist in every other state, too, doing their part to help keep forests, prairies, coastlines, and other areas intact.

Interest in the United States really took off in the 1960s and 1970s, when awareness and concern about the environment started to grow. As of 2005, there were 1,667 private nonprofit land trusts, ranging from small, one- or two-person trusts to large statewide groups with paid staffs of 30 or more people.

There are also several large national land trust organizations that do land trust work themselves or provide support services to other land trusts. One is The Nature Conservancy (TNC), based in Arlington, Virginia, with state chapters nationwide. Established in 1951, today it employs contract and seasonal workers and emphasizes conservation of "rare or relatively rare" species and natural communities.

Another key group is the Trust for Public Land (TPL) in San Francisco, established in 1972. An early TPL success was buying up miles of San Francisco coastline, therefore rescuing it from developers' hands. (The National Park Service now manages these areas.) Today, TPL also provides a wide range of services to other land trusts, from an informational newsletter to help with handling land transactions.

The Land Trust Alliance was set up in 1982 by trusts nationwide that wanted a central organization in Washington, D.C. In addition to providing information services, publications, documents, case studies, and other support to land trusts, it has a lobbyist to give land trusts a presence on Capitol Hill.

THE JOB

Depending on the land trust, land acquisition may be one person's entire job, or it may be one task among many for an executive director or other employee. Larger, well-funded land trusts and national land trust organizations are most likely to have acquisitions professionals devoted solely to handling land transactions. In smaller organizations, one person may do everything from land acquisition to fund-raising.

In any case, a group wishing to save some land or water has important questions to answer. Who owns it? Would they be willing to donate the land? If not, will they sell it, and if so, for how much? Who'll pay for it? Will a public agency buy it? Could a community group raise the needed funds?

A land trust can check with the local government to see if it is interested in helping to acquire the land. If that does not work, the trust can act on its own, trying to talk the landowner into selling or donating the land, for example. Or it can turn to big groups like the Land Trust Alliance, TNC, or TPL for help. At TPL, for example, project managers are available to help coordinate acquisitions efforts for other land trusts.

Generally, acquisitions involve either buying the land outright, acquiring development rights to it, obtaining easements, or getting landowners to donate the land or leave it to the land trust in their wills. Such negotiations usually do not go through the courts, but they do go through a type of legal process that is similar in some ways to buying a house. This usually involves having the land appraised and getting the title or deed to the property. After these tasks are accomplished, the land then becomes the property of the land trust.

For land donations, nonprofits offer certain advantages to landowners over donating to government or quasi-government groups. Generally, anything donated to a nonprofit is tax deductible. Large landowners may gain certain additional tax benefits by willing the land to a nonprofit when the landowner dies.

In general, most government agencies are not set up to receive donations of land. Landowners also may like the nonprofits' conservation emphasis and may not like the idea of donating their land to the government.

Instead of selling, donating, or willing the land, landowners might instead agree to easements that effectively put part of the land off-limits to development, subdivision, or other actions that might threaten preservation. For example, a large paper company gave

TNC agricultural and environmental easements on property the paper company owns near Richmond, Virginia, on which stands the oldest working farm in America. As with a land donation, the farmer or landowner who agrees to the easement gets some kind of tax break. In addition to the tax break, the landowner has the pleasure of doing something for the environment as well.

How does the land trust decide what land or water it wants to try to save? That varies widely. Sometimes it is a matter of wanting to make sure there is park area in a new residential development. Sometimes the issues are larger. TNC, for example, emphasizes acquiring areas where there is a threat to a natural community. This may involve endangered species, but TNC also thinks in terms of "rare" and "relatively rare" species, and of the uniqueness of the land—saving areas representing the best of their kind, such as the best oak hardwood forests, for example. Databases help keep track of such efforts.

REQUIREMENTS
High School
Many courses available to high school students can serve as a good preparation for a career as a land acquisition professional. Science courses from biology to ecology will help you understand the technical aspects of environmental concerns and appreciate the lands you'll be working with. Communications courses, such as English and speech, will help you negotiate with landowners, while business and math classes will prepare you for work with contracts and tax documents.

Postsecondary Training
A bachelor's degree is usually a solid foundation for this type of work. Good negotiating and deal-making skills are critical to land acquisition professionals. Land trust and preserve work in general draws people from all kinds of career areas, from city planning and land-use consulting to law and journalism. Real estate backgrounds are especially useful for people wishing to concentrate on the acquisitions side of land trust work.

Other Requirements
Communication skills are very important. Land acquisition professionals need to be able to think on their feet and have two or three alternative scenarios in mind when they meet to negotiate.

EXPLORING

To explore a career with land trusts, check your local library for land trust publications like the Land Trust Alliance's book *Conservancy: The Land Trust Movement in America*. You can also contact the large national land trust organizations for career information. These organizations should also be able to provide you with the names of local groups to get involved with.

STARTING OUT

Volunteering for or doing an internship with a land trust is an excellent way to enter the field. Large statewide organizations are probably the best bets for internships, as are the national organizations. Finally, one option for entering this field is to help start a land trust in your area yourself!

ADVANCEMENT

Advancement will depend on the size of the land trust organization. A project manager with the Trust for Public Land might move up into an administrative position, for example, or get more complicated cases. Other options might be to move over to a federal agency that manages federal lands, although these jobs are scarce, or into

The Whys of Land Trusts

Why do trusts acquire sites?

- To stop overly heavy grazing, farming, or recreation on the land
- To keep open lands from being bought by a developer
- To clean up a pond or lake and bring back native wildlife
- To manage lands with rare or endangered species

Why do owners turn land over to trusts?

- To do something good for the environment
- Historic preservation
- To obtain a tax break

the for-profit sector, such as with a consulting firm or a private company that manages large parcels of land, like a timber company.

EARNINGS

Less than half of the land trusts have paid staff. However, executive directors of land trusts may earn salaries that range from $21,860 to $102,250 annually, depending on the size of the trust.

Full-time land trust and preserve managers usually receive fringe benefits such as paid vacation and sick days, health insurance, and 401(k) plans.

WORK ENVIRONMENT

This job can take a land acquisition specialist literally "out in the field" to check out land or water parcels the land trust is considering acquiring; it also can bring them to the negotiating table, where they will be sitting down with landowners hashing out a deal. Land trusts exist all over the country. Work with a large national organization might involve travel to help out the smaller land trust organizations. Hours, benefits, and other particulars will vary depending on the specific land trust and its resources.

OUTLOOK

The outlook for land trust work currently is brighter than that for federal land and water conservation jobs. Land trusts are going strong right now, and the entire land and water conservation segment, of which land trust and preserve management is a part, is growing at a steady rate annually. The Land Trust Alliance reports that the number of land trusts increased by 32 percent from 2000 to 2005.

FOR MORE INFORMATION

The following is a national organization of more than 1,700 land trusts nationwide:

Land Trust Alliance
1660 L Street, NW, Suite 1100
Washington, DC 20036-5635
Tel: 202-638-4725
Email: info@lta.org
http://www.landtrustalliance.org

The following organization specializes in land trusts and land trust management for areas with rare or endangered species. For information about internships with TNC state chapters or at the TNC headquarters, contact

The Nature Conservancy (TNC)
4245 North Fairfax Drive, Suite 100
Arlington, VA 22203-1606
Tel: 703-841-5300
http://nature.org

For information on opportunities for students, contact

Student Conservation Association
PO Box 550
689 River Road
Charlestown, NH 03603-0550
Tel: 603-543-1700
http://www.thesca.org

For information on land conservation careers, contact

The Trust for Public Land
116 New Montgomery Street, Fourth Floor
San Francisco, CA 94105-3638
Tel: 415-495-4014
Email: info@tpl.org
http://www.tpl.org

Landscape Architects

OVERVIEW

Landscape architects plan and design areas such as highways, housing communities, college campuses, commercial centers, recreation facilities, and nature conservation areas. They work to balance beauty and function in developed outdoor areas. There are approximately 28,000 landscape architects employed in the United States.

HISTORY

In the United States, landscape architecture has been practiced as a profession for the last 100 years. During the early part of the 20th century, landscape architects were employed mainly by the wealthy or by the government on public works projects. In 1918 the practice of dividing large plots of land into individual lots for sale was born. In addition, there was a new public interest in the development of outdoor recreational facilities. These two factors provided many new opportunities for landscape architects.

The most dramatic growth occurred following the environmental movement of the 1960s, when public respect for protection of valuable natural resources reached an all-time high. Landscape architects have played a key role in encouraging the protection of natural resources while providing for the increasing housing and recreation needs of the American public.

In the last 30 years, the development of recreational areas has become more important, as has the development of streets, bypasses, and massive highways. In addition, the increased focus on creating functional, aesthetic designs that are in harmony with the environment has created greater demand for landscape architects. Both developers and community planners draw upon the services of landscape architects now more than ever.

THE JOB

Landscape architects plan and design outdoor spaces that make the best use of the land and at the same time respect the needs of the natural environment. They may be involved in a number of different types of projects, including the design of parks or gardens, scenic roads, housing projects, college or high school campuses, country clubs, cemeteries, or golf courses. They may also be involved in the restoration of mines or landfills. They may even be hired to help with "traffic calming"—slowing traffic through design and enhancement of the physical environment. Landscape architects can work in both the public and private sectors.

Landscape architects begin a project by meeting with the client to learn what type of environment they're seeking. They discuss the purpose of the site, structures needed, and the budget for the work. They then study the work site itself, observing and mapping such features as the slope of the land, existing structures, plants, and trees. They also consider different views of the location, taking note of shady and sunny areas, the structure of the soil, and existing utilities.

Landscape architects consult with a number of different people, such as engineers, architects, hydrologists, city officials, zoning experts, real estate agents and brokers, and landscape nursery workers to develop a complete understanding of the job, and to determine if the job is, in fact, feasible. Awareness of environmental and other regulations plays an important role in this work. Using computer-aided design (CAD) software, landscape architects create detailed plans and drawings of the site, factoring in topography and grading, to present to the client for approval. They may also present their ideas to clients by using video simulation, as well as geographic information systems (computer mapping) technology. They tackle such issues as calculating the amount of soil to be removed or added to the site, the placement and building of retaining walls, as well as

Landscape architects plant a drought-resistant grass native to south Louisiana on the Global Green's solar home rooftop deck. The grassy area will help insulate the house and collect rainwater that will be channeled to the home's cistern in New Orleans. Rather than a traditional pitched roof, this one slopes high on one side to allow the roof's solar panels to get the maximum amount of sun exposure. *AP Photo/Judi Bottoni*

the arrangement of roads and buildings. Some projects take many months before the proposed plans are ready to be presented to the client.

After developing final plans and drawing up a materials list, landscape architects invite construction companies to submit bids for the job. Depending upon the nature of the project and the contractual agreement, landscape architects may remain on the job to supervise construction, or they may leave the project once work has begun. Those who remain on the job serve as the client's representative until the job is completed and approved.

REQUIREMENTS
High School
Landscape architects rely on a variety of skills and talents in their work. To prepare for a college program in landscape architecture, take courses in English composition and literature; art; social sciences, including history, government, and sociology; natural sci-

ences, including biology, chemistry, and physics; and mathematics. If available, take drafting and mechanical drawing courses to begin acquiring the technical skills needed for the career.

Postsecondary Training

A bachelor's or master's degree in landscape architecture is usually the minimum requirement for entry into this field. Undergraduate and graduate programs in landscape architecture are offered in various colleges and universities. In 2007, the Landscape Architectural Accreditation Board of the American Society of Landscape Architects (ASLA) accredited 79 programs at 61 colleges and universities. Courses of study usually focus on: landscape design and construction, plant and soil science, landscape ecology, architecture, urban and regional planning, surveying, graphic expression (mechanical, freehand, and computer-based drawings), and professional practice and general management.

Hands-on work is a crucial element to the curriculum. Whenever possible, students work on real projects to gain experience with computer-aided design programs and video simulation.

Certification or Licensing

Almost all states require landscape architects to be licensed. To obtain licensure, applicants must pass the Landscape Architect Registration Examination, sponsored by the Council of Landscape Architectural Registration Boards (CLARB). Though standards vary by state, most require applicants to have a degree from an accredited program and to be working toward one to four years of experience in the field. In addition, 15 states require prospective landscape architects to pass another exam that tests knowledge of local environmental regulations, vegetation, and other characteristics unique to the particular state. Because these standards vary, landscape architects may have to reapply for licensure if they plan to work in a different state. However, in many cases, workers who meet the national standards and have passed the exam may be granted the right to work elsewhere. For more information on licensing, contact the CLARB (http://www.clarb.org) or the ASLA (http://www.asla.org).

Landscape architects working for the federal government need a bachelor's or master's degree but do not need to be licensed.

Other Requirements

You should be interested in art and nature and have good business sense, especially if you hope to work independently. Interest in environmental protection, community improvement, and landscape design is also crucial for the profession. You should also be flexible and be able to think creatively to solve unexpected problems that may develop during the course of a project.

EXPLORING

If you are interested in learning more about the field, you can gather information and experience in a number of ways. Apply for a summer internship with a landscape architectural firm or at least arrange to talk with someone in the job. Ask them questions about their daily duties, the job's advantages and disadvantages, and recommendations for any landscape architecture programs. Visit the Career Discovery section of the ASLA Web site to read more about the job (http://www.asla.org/CareerDiscovery.aspx). And to further gauge your interest in the field, take ASLA's Landscape Architecture Interest Test (http://archives.asla.org/nonmembers/recruitment/lainttest.htm).

EMPLOYERS

There are roughly 28,000 landscape architects employed in the United States, and about 19 percent of them are self-employed. Landscape architects are found in every state in the United States, in small towns and cities as well as heavily populated areas. Some work in rural areas, planning and designing parks and recreational areas. Many work in suburban and urban areas, however, where the majority of positions are found.

A variety of different employers in both the public and private sectors hire landscape architects to work on projects. They may work with a school board planning a new elementary or high school, with manufacturers developing a new factory, with homeowners improving the land surrounding their home, or with a city council planning a new suburban development. Many are also being called on to help with historic landscape preservation or renovation.

In the private sector, most landscape architects do some residential work, though few limit themselves entirely to projects with individual homeowners. Larger commercial or community projects are usually more profitable. Workers in the public sector plan and design government buildings, parks, and public lands. They also may conduct studies on environmental issues and restore degraded lands.

STARTING OUT

After graduating from a landscape architecture program, you can usually receive job assistance from the school's career placement service. Although these services do not guarantee a job, they can be of great help in making initial contacts. Many positions are posted by the American Society of Landscape Architects and published in its two journals, *Landscape Architectural News Digest Online* (http://land.asla.org) and *Landscape Architecture* (http://archives.asla.org/nonmembers/lam.html). Government positions are normally filled through civil service examinations. Information regarding vacancies may be obtained through the local, state, or federal civil service commissions.

Most new hires are often referred to as interns or apprentices until they have gained initial experience in the field and have passed the necessary examinations. Apprentices' duties vary by employer; some handle background project research, others are directly involved in planning and design. Whatever their involvement, all new hires work under the direct supervision of a licensed landscape architect. All drawings and plans must be signed and sealed by the licensed supervisor for legal purposes.

ADVANCEMENT

After obtaining licensure and gaining work experience in all phases of a project's development, landscape architects can become project managers, responsible for overseeing the entire project and meet-

Landscaping and GIS

The use of satellite imagery and GIS (geographic information systems) in landscape architecture has transformed the field. Using this technology, workers can study a potential work site from a different location. For example, landscape architects can use remote sensors to analyze the overall condition of a wetland. This information is then analyzed to determine how a new proposed freeway will affect the area. Satellite imagery and GIS enable the landscape architect to study environmental relationships and better understand ecosystems, environmental disasters, and land use patterns.

ing schedule deadlines and budgets. They can also advance to the level of associate, increasing their earning opportunities by gaining a profitable stake in a firm.

The ultimate objective of many landscape architects is to gain the experience necessary to organize and open their own firm. According to the U.S. Department of Labor, approximately 19 percent of all landscape architects are self-employed—more than two times the average of workers in other professions. After the initial investment in computer-aided design software, few start-up costs are involved in breaking into the business independently.

EARNINGS

Salaries for landscape architects vary depending on the employer, work experience, location, and whether they are paid a straight salary or earn a percentage of a firm's profits.

According to 2008 data from the U.S. Department of Labor, the median annual salary for landscape architects was $58,960. The lowest paid 10 percent earned less than $36,520 and the highest paid 10 percent earned more than $97,370. Landscape architects who worked for architectural, engineering, or other related services in 2008 earned median annual salaries of $64,700. Those who worked for the federal government in this same year earned median incomes of $80,830, while those who worked for lawn and garden suppliers and stores had lower annual salaries, averaging $53,120 per year.

Benefits also vary depending on the employer but usually include health insurance coverage, paid vacation time, and sick leave. Many landscape architects work for small landscaping firms or are self-employed. These workers generally receive fewer benefits than those who work for large organizations.

WORK ENVIRONMENT

Landscape architects spend much of their time in the field gathering information at the work site. They also spend time in the office, drafting plans and designs. Those working for larger organizations may have to travel farther away to work sites.

Work hours are generally regular, except during periods of increased business or when nearing a project deadline. Hours vary for self-employed workers because they determine their own schedules.

Landscape Architecture and Public Health

When planning a residential community, landscape architects must consider a number of factors, including the nature of the land available for the project, budget constraints for the project, and the needs of their clients. However, there is another, very important factor that more and more landscape architects have begun to consider: the impact of the community on public health.

Statistics on Americans' health have been increasingly negative in recent years; for example, incidence of obesity, heart disease, and diabetes have been on the rise. Many health professionals attribute these trends to the increasingly sedentary, inactive lifestyles that many Americans lead. This is due, in part, to community design. For example, in many parts of the country, simply walking from one's home to the grocery store is impractical, if not impossible. In other communities, the development of sidewalks and outdoor public spaces may have given way to the development of more homes, thus eliminating opportunities for people to enjoy basic physical activity in their neighborhoods.

Many landscape architects have begun working on healthy, "walkable" community designs. Some of the main features of such designs are as follows:

- Safe and accessible sidewalks, crosswalks, and bike paths
- Easy access to mass transportation
- Safe and attractive public parks and facilities
- Shopping and service centers that people can reach by walking

Source: American Society of Landscape Architects

OUTLOOK

According to the *Occupational Outlook Handbook,* the employment of landscape architects is expected to increase faster than the average for all occupations through 2016. The increase in demand for landscape architects is a result of several factors: a boom in the construction industry, the need to refurbish existing sites, and the increase in city and environmental planning and historic preservation. More landscape architects will also be needed to help build and restore sites that meet environmental regulations and coding while having as little impact as possible on the environment. In addition, many job openings are expected to result from the need to replace experienced workers who leave the field.

The need for landscape architecture depends to a great extent on the construction industry. In the event of an economic downturn, when real estate transactions and the construction business is expected to drop off, opportunities for landscape architects will also dwindle.

Opportunities will be the greatest for workers who develop strong technical skills. The growing use of technology such as computer-aided design will not diminish the demand for landscape architects. New and improved techniques will be used to create better designs more efficiently rather than reduce the number of workers needed to do the job.

FOR MORE INFORMATION

For information on the career, accredited education programs, licensure requirements, and available publications, contact

American Society of Landscape Architects
636 Eye Street, NW
Washington, DC 20001-3736
Tel: 202-898-2444
http://www.asla.org

For information on student resources, license examinations, and continuing education, contact

Council of Landscape Architectural Registration Boards
3949 Pender Drive, Suite 120
Fairfax, VA 22030-6088
Tel: 571-432-0332
Email: Info@Clarb.org
http://www.clarb.org

For career and educational information, visit the following Web site sponsored by the Landscape Architecture Foundation:

LAprofession.org
http://www.laprofession.org

Land Trust or Preserve Managers

OVERVIEW

Land trust or preserve managers are part of private and federal efforts to preserve land or water from development; subdivision; pollution; overly heavy recreational, grazing, agricultural, or other use; or other human action. The management tasks of land trusts or preserves vary widely, from monitoring the site, inventorying species, or managing natural resources via specialized conservation and preservation work. Examples of the latter might include doing controlled burnings, re-creating lost or damaged ecosystems, and restoring native plants and animals.

HISTORY

Efforts to conserve land and water date back more than 100 years in this country and have been driven by two key forces: the government and private citizens' or community groups. Alarm about diminishing wilderness areas in the West led to the establishment of the first national parks and preserves by our government in the late

19th century. Around that time, the government also set aside four Civil War battlefields as national battlefield parks, the first historic sites so acquired by the U.S. government.

The single most influential figure in early conservation efforts was Theodore Roosevelt, the 26th president of the United States. Roosevelt fell in love with the West as a young man, when ill health led him there to seek better air. He owned a ranch in the Dakota Territory and wrote many books about his experiences in the West.

When he became president in 1901, Roosevelt used the position to help preserve his beloved West. He and his administrators pushed conservation as part of an overall strategy for the responsible use of natural resources, including forests, pastures, fish, game, soil, and minerals. This both increased public awareness of and support for conservation and led to important early conservation legislation. Roosevelt's administration especially emphasized the preservation of forests, wildlife, parklands, wilderness areas, and watershed areas and carried out such work as the first inventory of natural resources in this country.

Roosevelt was very proud of the monumental accomplishments of his administration in conserving the natural resources of the nation. He wrote, "During the seven and one-half years closing on March 4, 1909 (the years of his administration), more was accomplished for the protection of wildlife in the United States than during all the previous years, excepting only the creation of Yellowstone National Park."

But government action is only part of this story. Individual citizens forming private nonprofit land trusts, plus national nonprofit land trust organizations, have saved countless acres of land and water as well. They, too, have their roots in the last century.

Back in 1891, the city of Boston was bursting at the seams. A thriving shipbuilding industry plus other commercial and industrial pursuits had helped that city boom in the 19th century. Boston also had seen an explosion in immigrant population, particularly Irish immigrants. The captains of the industry and their families poured money into the arts, helping Boston gain a reputation as the "Athens of America."

Some Bostonians, however, were troubled by the rapid development that swallowed up areas at the edges of the city. They were concerned that remaining wild areas were going to disappear and that many people living in the city were never going to have access to open lands and wilderness areas.

One group of citizens took action. They formed a group called the Trustees of Reservations, bought up some of the undeveloped land themselves, and opened the areas to the public for recreational use. This was the first official land trust in the country, and it paved the way for a whole movement of private land trusts.

Individuals as well as large groups have started land trusts; they have worked to protect as little as a few acres of land to as much as hundreds of acres, depending on the part of the country and the trust's resources. Sometimes trusts just acquire the land or easements on it; but sometimes, and increasingly in recent years, they also take steps to environmentally manage it.

Land trusts saw very strong growth in the mid- to late-1980s. Following a slight dip in the early 1990s due to a recession, they are going strong today. In 2005 there were 1,667 private nonprofit land trusts, up nearly four times that of the number of trusts (450) that existed back in 1982. And according to the Land Trust Alliance, an organization based in Washington, D.C., consisting of 1.5 million land trust supporters and members, the total number of acres protected by local, state, and national land trusts doubled to 37 million acres between the years 2000 and 2005.

Sometimes land trusts work in cooperation with U.S. federal land-management agencies. This is true of The Nature Conservancy (TNC), for example, a very large national land trust organization specializing in rare wildlife and habitats. The organization assists agencies such as the Environmental Protection Agency, Department of Agriculture, U.S. Agency for International Development, Department of the Interior, National Park Service, and Department of Defense with the management of land and biodiversity conservation. The conservancy also works with state and local governments, nonprofit organizations, corporations, and private individuals.

Consulting firms specializing in land trust or preserve management also exist and may be called in to help with special areas like ecosystem restoration or forestry management. Finally, some private corporations, such as utility companies or timber companies, own and manage large parcels of land; their land management may include conservation and preservation of areas such as forest wetlands.

THE JOB

Land trusts acquire land by buying it, getting the landowner to donate it, arranging for easements on it, or purchasing the development rights to it. Land acquisition may be just one of many tasks of

Dr. Michael Looker, left, with Trust for Nature, and Malcolm Tonkin, manager of environmental services with Hancock Victorian Plantations, work to keep the forest healthy in Latrobe, Victoria, Australia. Environmentalists and loggers are working together to manage an old-growth forest, balancing logging with protection of the forest. *Getty Images*

a land trust employee, such as the executive director; or, in larger land trusts, it may be the sole job of one or more land acquisition professionals.

The duties of managing a land trust or preserve depend on the specific land or water involved and its needs, who is doing the managing, how much funding and staffing is available, and other factors.

Staffing of land trusts can be minimal, particularly in the early years of the trust. At first, one person might do everything from handling correspondence to walking the land. If the land trust grows larger, it may add more people who can then focus on specific tasks, including management of the land, fundraising operations, events planning, membership outreach, and more. A few land trusts, particularly some of the large statewide land trusts, are large enough to have a staff of 30 paid people or more.

As for federally managed lands, these, too, can have varying levels of staffing and funding that affect what specific work is done. But the federal government employs about 75 percent of all people working in land and water conservation; in general the federal agencies have greater resources than private land trusts. For example, all national parks have natural resource management departments that carry out tasks from ensuring environmental compliance to specialized conservation or preservation work.

Specific work varies in different parts of the country, from Eastern forests to the Everglades to coastal areas, and ranges from simply monitoring the land to doing specialized work like re-creating destroyed ecosystems. Examples include:

Planning for better use of land and water. If the land is a recreational area, for example, managers might plan how to prevent overuse.

Species inventory. Cataloging plant and animal species helps establish the baseline needed to create short- and long-term plans for the land.

Restoration or re-creation of damaged or destroyed ecosystems. Getting an area back to how it used to be may involve cleaning up pollution, bringing back native species, and getting rid of non-native species. Landscape architects, biologists, botanists, ecologists, and others may help do such work. Restoration of wetlands, one example of this work, including forest wetlands, may involve wetlands ecologists, fish and wildlife scientists, and botanists.

Habitat protection. Protecting wildlife habitats, particularly those of rare or endangered species, is another important task. At least 1,317 plants and animals in the United States alone currently are endangered or threatened, according to the U.S. Fish and Wildlife Service.

Prescribed burnings. Management of prairies, forests, or rangelands may involve controlled burnings. After the fire, specialists may go in and inventory species. Pitch pine communities in New York and New Jersey, and long-leaf pine forests in Virginia, Texas, and other parts of the South, are some areas handled in this way.

Rangeland management. In addition to prescribed burnings, this may involve controlled grazing by bison or cattle to keep plant life under control.

REQUIREMENTS
High School
Recommended high school course work for those interested in scientific work includes biology, chemistry, and physics as well as botany and ecology. All potential land trust or preserve managers can benefit from courses in business, computer science, English, and speech.

Postsecondary Training
At the undergraduate level, you might get a degree in one of the natural sciences, such as zoology, biology, or botany. There has also been growing interest in degrees in conservation biology, which focuses on the conservation of specific plant and animal communities, from schools such as the University of Wisconsin—Madison (http://www.wisc.edu). Another key program is the School of Forestry and Environmental Studies (http://environment.yale.edu) at Yale University. Land and water conservation is a popular field, so if you are interested in the natural sciences, you are advised to earn at least a master's degree.

Other Requirements
To be successful in this field, according to Robert Linck, Director of Conservation Funding at the Vermont Land Trust in Montpelier, Vermont, "you must be hard working and dedicated to the field of land conservation, loyal to the organization, have good people skills, and the ability to speak and write clearly. You must also be efficient and able to juggle multiple responsibilities in a reasonably organized fashion and have knowledge of land conservation options and techniques, a willingness to participate in fund-raising, and a good sense of humor."

Because land trusts tend to be entrepreneurial, you will also need to be skilled in business administration, finance, and law—especially if your duties involve running the financial end of the trust, raising funds, negotiating deals, and handling tax matters.

EXPLORING

There are many ways to explore a career in land and water conservation. Read up on land and water conservation in the library, contact nonprofit land trusts or federal agencies for information about current projects, or check out the degree programs at local universities. The Internet is another rich source of up-to-date information; some useful sites are listed at the end of this article.

EMPLOYERS

The federal government, in its various agencies and branches, is the largest employer of land trust professionals. State and local government agencies also employ some land trust professionals in a variety of positions. Outside of government, potential employers include numerous nonprofit organizations and private land trusts. Additionally, large banks and other similar institutions employ land trust specialists.

STARTING OUT

This field is so popular that many people get their start in less traditional ways, such as through contract or seasonal work, volunteer work, and internships.

Even people graduating with a master's degree may only be able to land contract work at first, which is work done on a per-project or freelance basis. Contract workers usually are specialists, such as ecologists or botanists. The need for them is high in the summer months when biological inventorying work is plentiful.

"The best way to learn about opportunities in the land trust community," according to Robert Linck, "is through the Land Trust Alliance (LTA), which maintains a Web site that includes job listings. LTA also has a listserv that often has postings for position openings. The Nature Conservancy and the Trust for Public Land also maintain Web sites with job listings. Direct contact with land conservation organizations in any particular region could also inform a job seeker of current or anticipated opportunities."

Volunteer and internship opportunities are available at many environmental organizations. These opportunities frequently lead to paid positions and always provide valuable experience.

ADVANCEMENT

There are three general advancement paths for land trust professionals. The traditional promotion path might begin with an internship, then progress to positions of increasing power and responsibility. The second path involves expansion of duties within a specialty field. For example, someone who starts out as a land protection specialist in North Carolina may not have any desire to move out of that work; therefore, his or her job may be expanded laterally—broadening into consulting work in the specialty in other parts of the state, or even nationwide. Third, a person may opt for a demotion—getting back to land protection and conservation fieldwork, for example, after having served in an administrative position.

EARNINGS

According to the U.S. Department of Labor, graduates with a bachelor's degree who worked for the federal government received average starting salary offers of between $28,862 and $35,752 in 2006, depending on academic achievement. Salaries for conservation workers ranged from less than $35,190 to $86,910 or more annually.

The National Association of Colleges and Employers cites that graduates with a bachelor's degree in conservation and renewable natural resources received an average starting salary offer of $34,678 in July 2007.

Federal government agency jobs pay more than state or local government jobs. Nonprofit groups' salaries can be competitive but tend to be at the lower end of the pay range. Salaries also tend to vary by region.

WORK ENVIRONMENT

Tramping around in the wilderness, inventorying plant and animal species, working outdoors to help develop a natural area—all of these are possibilities for people working in land or water conservation, particularly if they are working as a natural scientist or in

(continues on page 110)

INTERVIEW

Robert Linck is the director of conservation funding for the Vermont Land Trust (VLT) in Montpelier, Vermont.

Q. What do you do in your job?

A. I work with regional directors to pick projects that need extra funding. We'll choose two to four, and then talk to people to raise the money. We look at foundation sources, and all the ways it can take to raise money. I work with field team members, and I meet with supporters, gauge their interest, and solicit gifts for specific projects as well as for general support. Fundraising and foundation work is a work in progress. We have tapped less private funding because we've been fortunate with our funding sources. I'll be helping to beef up the cultivation of foundations. With the new [Obama] administration, we will see more resources go to conservation, so we'll be working on this.

Q. What did you do in your previous position?

A. I co-directed the land conservation activities of the Vermont Land Trust's Champlain Valley Region and helped manage the staff, operations, and budget for a six-person office. I assessed the conservation values of land according to Land Trust criteria; negotiated or managed negotiations with landowners in conservation transactions; raised funds for conservation transactions from foundations, state and federal sources, communities, and individual donors; conducted outreach to landowners, communities, other nonprofit organizations, and the general public; represented the Vermont Land Trust before town boards, state agencies, the legislature, and the media; worked with other staff on organizational policy and systems issues; and assisted headquarters staff with annual and spring fund-raising appeals and membership events.

Q. What do you like most about your work?

A. I like meeting people, learning what interests them in conservation, and what brought them into supporting the organization. We do a lot of work with farmers and with the farm economy—helping to determine the best soils to be farmed and making sure the land is used well. We work with communities on things that are important to them, and on things that are important to society. We touch on that all the time. Getting transactions completed is satisfying.

Q. What is your work environment like?

A. I work with dedicated coworkers in a positive and relatively fast-paced environment in a well-run nonprofit organization. Our office is in a small town in a rural Vermont setting.

Q. What has surprised you most about this work?

A. The most surprising thing about the work is how deeply satisfying it is. I've been interested in this since the early 1970s, and at it as a career for 30 years. The results are so tangible: you conserve land. Our work is permanent. We can see the benefits. The public can see the benefits. Another surprising thing, though not a positive one, is that you don't get outside as much as you might think. You interact with legal folks, and sit in lots of meetings and work on the computer. You may visit the property two or three times, but that can be it for outside work.

Q. What expectations did you have when you first entered this field?

A. I suppose I had fairly clear expectations when I entered land conservation work, in part because one of my college professors was associated with a local land trust. My first job involved a broad range of responsibilities, including land conservation, so I immediately immersed myself in the kind of work that I have come back to years later. Perhaps one misconception was laid to rest early on—most land trust employees do not spend vast amounts of their time outdoors. The other thing to note is that, perhaps like most jobs, much of what you learn happens on the job.

Q. What course of undergraduate and graduate study did you pursue? Did this prepare you for your career? Also, did you complete any internships or special training for this career?

(continues)

(continued)

A. I have a B.S. in environmental studies and biology and an M.S. in water resources management. For five years in between those degrees I was employed by a nonprofit watershed organization. I participated in four internships during college, but none of those was oriented toward land conservation. Otherwise, the training for my current position has resulted from previous job experience and from workshops, seminars, and annual conferences of land conservation professionals organized by the Land Trust Alliance. The Vermont Land Trust also conducts very effective staff training, as needed. My educational background was very instrumental in my career, but the experiences of and relationships built through several internships and previous jobs were equally important.

Q. What other types of positions have you held?
A. I was regional (Vermont/New Hampshire) director for a four-state watershed organization—the Connecticut River Watershed Council; conservation director for the Adirondack Mountain Club; recycling coordinator for Warren County, New York; an adjunct professor of environmental studies at Adirondack Community College; an extension specialist for the Hudson River Estuary, New York Sea Grant/Cornell University; associate director for the Upper Valley Land Trust, Hanover, New Hampshire; and executive director of the Southeast Arizona Land Trust.

Q. What is the current outlook for growth and advancement in your field?
A. The outlook is very promising. Over 1,600 land trusts operate around the United States today, and the numbers are still growing. Though some of them are very small and may involve only volunteers, many more are already large organizations or are growing, so land conservation continues to have a strong future. We won't be doing land conservation forever. There

(continued from page 107)
support of the scientists. Administrators, communicators, lawyers, and others more often will find themselves in offices, of course, especially when they are working for larger organizations. Outdoor

will be a saturation point in different areas. There will come a point when people will say, "When are we going to stop doing this here?" There will be fewer projects done, but land trusts are perpetual. You are fund-raising to draw money for endorsements. You may not own the land, but you maintain relationships with the landowners, and you continue to visit the land. We're endowed to do this. We'll be doing this forever, for however long that is. So it's a secure job, in that sense, for people who are doing this type of work.

Q. What advice would you give to someone who is interested in pursuing a career in land trust or preserve management?

A. Fully understand the skill set required for the position you desire, take advantage of training opportunities, seek college- and graduate-level education in existing or emerging programs that emphasize land conservation or a key skill associated with land conservation, and pursue opportunities to volunteer or intern at a land trust.

Many land trust jobs involve the type of work described above. Many other positions are focused on management and administration, geographic information systems (GIS) and mapping, fund-raising and community relations, or legal/paralegal work. For land trusts that own land and emphasize ownership and management of land or preserves, some positions will involve much more work outdoors, "in the field." Land trusts (including the Vermont Land Trust) that emphasize the stewardship of conservation easements have positions oriented toward landowner relations, "baseline documentation" of properties that have been conserved (GIS and mapping, field work with maps, GPS operation), field monitoring of private or public conserved land, interpretation of legal documents, and managing a legal process when conservation easement violations occur.

jobs in this field are very popular. The Environmental Conservation Organization says people in land and water conservation tend to stay in their jobs longer than people in other environmental careers, attesting to the appeal of these jobs.

OUTLOOK

The best opportunities in this field are with the private land trusts and national land trust organizations, as opposed to the federal agencies. With little exception, none of the federal agencies is expected to see big growth over the next few years. On the other hand, following the slight slowdowns of the early and late 1990s, the private land trusts are growing.

Land trusts are the fastest growing arm of the conservation movement today, with approximately 1,667 in 2005, according to the Land Trust Alliance. LTA's National Land Trust Census reports that local and regional land trusts protected nearly 11.9 million acres as of 2005—more than double the amount of acreage protected in 1998.

FOR MORE INFORMATION

The following is a national organization of more than 1,700 land trusts nationwide:

Land Trust Alliance
1600 L Street, NW, Suite 1100
Washington, DC 20036-5635
Tel: 202-638-4725
Email: info@lta.org
http://www.landtrustalliance.org

This conservation organization offers fellowships for graduate work in conservation, places people in paid internships, and more.

National Wildlife Federation
11100 Wildlife Center Drive
Reston, VA 20190-5362
Tel: 800-822-9919
http://www.nwf.org

The following organization specializes in land trusts and land trust management for areas with rare or endangered species. For information about internships with TNC state chapters or at the TNC headquarters, contact

The Nature Conservancy (TNC)
4245 North Fairfax Drive, Suite 100
Arlington, VA 22203-1606
Tel: 703-841-5300
http://www.nature.org

Contact this group for information regarding volunteer positions in natural resource management, including with federal land management agencies.

Student Conservation Association
PO Box 550
689 River Road
Charlestown, NH 03603-0550
Tel: 603-543-1700
http://www.thesca.org

For information on land conservation careers, contact

The Trust for Public Land
116 New Montgomery Street, Fourth Floor
San Francisco, CA 94105-3638
Tel: 415-495-4014
Email: info@tpl.org
http://www.tpl.org

Managers and Owners, Green Business

QUICK FACTS

School Subjects
Business
Communications
English

Personal Skills
Analytical
Creative
Decisive

Work Environment
Primarily indoors
Primarily one location

Minimum Education Level
Bachelor's degree

Salary Range
$27,770 to $68,680 to
$166,400+

Certification or Licensing
Not required

Outlook
About as fast as the
average

OVERVIEW

Managers and owners of green businesses own and/or manage companies that offer products or services focused on improving the environment. Green businesses can include companies that make products from recycled materials; build homes, factories, and other structures using organic and natural products; or offer products (such as solar, renewable-energy, and eco-products) that solve environmental problems.

HISTORY

In the 1960s people started paying more attention to harm being done to the environment and all living creatures by pollution and litter. Rachel Carson's book *Silent Spring*, published in 1962, raised awareness even further about human beings' impact on the environment by pointing out the dangers of insecticides, weed killers, and other agricultural sprays, explaining that people were being exposed to poisons that stayed in their bodies their entire lives.

To address these and many other issues, the U.S. Environmental Protection Agency was formed in 1970, with a mission to "protect human health and to safeguard the natural environment—air, water, and land—upon which life depends." In that same year, the National Environmental Policy Act was signed, requiring all federal agencies to conduct thorough assessments of environmental impact of all major programs. Also, the first Earth Day occurred on April 22, 1970. Thousands of new environmental laws and ordinances were enacted throughout the 1970s, including the Clean Air Act, Safe Drinking Water Act, Federal Water Pollution Control Act, Consumer Product Safety Act, Environmental Pesticide Control Act, and Endangered Species Act.

The idea of preventing pollution, rather than cleaning up after having caused it, came about in the late 1970s, pioneered by 3M (the Post-It Notes and Scotchgard manufacturer). By the 1980s, a growing number of companies started following 3M's example by changing their manufacturing and operating systems to use energy more efficiently and reduce costs. Environmental management systems came about in the 1990s. Companies realized that their success in "cleaning up their acts" depended heavily on their suppliers, contractors, and business partners, so they took more time to consider who to team up with on projects and plans, and what types of products to offer to consumers.

According to Joel Makower in his book *Strategies for the Green Environment*, the evolution of green business has had three stages, beginning with the Hippocratic oath, "First, do no harm;" followed by "Doing well by doing good," meaning companies that take steps toward corporate responsibility can improve their reputations; and the final stage was "Green is green," coined by General Electric Chairman Jeffrey Immelt, which means that companies that operate sustainably improve their finances.

THE JOB

Global concern about the impact of business on the environment has created greater demand for sustainable business and manufacturing practices, and green products. Green business managers and owners operate companies that meet these needs by creating products that do such things as improve energy efficiency, reuse and recycle materials, and work with, rather than against, nature. Their companies help improve communities and the environment and have missions to enhance quality of life.

Managers' and owners' duties and responsibilities vary, depending on the type and size of the company. In general, however, they help create company policies, manage and/or oversee daily functions, plan the use of equipment and materials, and oversee or run human resources.

Many business owners are entrepreneurs who come up with the idea for the company and start it themselves. They research the marketplace, confirm that their product fills a niche, and talk to and secure investors (such as venture capitalists, or angel investors) to help fund the start-up costs.

The origin of TOMS Shoes is a good example of a green company that sprang from one person's creative solution to a universal problem. In 2006 Blake Mycoskie, an American, was traveling in Argentina and befriended a group of children who had no shoes. He conceived of a shoe company with a "One for One" mission: for every pair of shoes purchased, TOMS would give a pair of shoes to a child in need. The shoes are basic and simple: lightweight, made from cotton or canvas, with rope or rubber soles, and the design is based on the espadrilles (known as "alpargata" shoes) Argentinean farmers have worn for decades. (The company also offers vegan styles.) Within the same year of founding TOMS, Blake, along with family, friends, and staff, returned to Argentina with 10,000 pairs of shoes that came from TOMS' customers. Since then, TOMS has given over 140,000 shoes to children through its One for One model, and aimed to give more than 300,000 shoes to children in need around the world in 2009.

In large companies, business managers usually manage a single department and oversee the work of their specific staff. If the company is large enough, the job may be even more defined, to the extent that some managers will oversee supervisors, reviewing job performance and reporting to upper-level management. Those who work for small companies are usually involved in more aspects of the company. They will have a wider variety of tasks and are often responsible for managing most, if not all, of the staff. They may also be involved in hiring employees and conducting regular employee reviews. Strategizing business plans, advertising campaigns, and promotion campaigns; creating employee and production schedules; and training employees can also be part of a manager's job description.

Retail business managers will oversee either a single department or the entire staff. They make sure all departments are communicating with each other and business goals are being met. They either oversee staff who handle the following responsibilities, or are directly

Dressed in her homemade "tree hugger" Halloween costume, Cassie Green, owner of Green Grocer Chicago, stands in her organic food market. *AP Photo/ Charles Rex Arbogast*

involved in these tasks themselves: stock inventory; merchandise inspection to make sure products are not outdated; pricing items and placing them on shelves and in displays; reviewing inventory and sales records; developing merchandise techniques; and meeting and greeting customers. They may also work on sales and promotion programs, prepare budgets, and approve sales contracts.

REQUIREMENTS
High School
Business owners and managers need to have strong organization skills to handle the day-to-day work, as well as clear communication skills to successfully manage and motivate their employees. Course work in business, math, English, and computers is a solid foundation for this type of work. A familiarity with environmental issues and science is also helpful, so be sure to take classes in biology and ecology.

Postsecondary Training
Most green business owners and managers have, at minimum, a bachelor's degree, which may be in liberal arts, business administration,

or science. While in college, take classes in business management, marketing, communications, environmental sciences, and environmental policy. Courses in sociology and psychology are also useful as a foundation for understanding people and what makes them tick.

A master's degree in business administration can help lower-level managers advance to positions of higher authority and greater pay, as well as provide business owners with a better understanding of business management.

Certification or Licensing

While not required, business managers can advance their careers by securing professional certification. Organizations such as the American Management Association and the Institute of Certified Professional Managers offer seminars and certification programs for supervisors, managers, and executives in all industries. Algonquin College in Canada also offers the Algonquin Green Business Management Certification Program to individuals who want to learn more about the best green practices.

Other Requirements

Green business owners and managers need to keep up with environmental laws and policies, green products and business trends, sustainable business practices, as well as general business management policies and procedures. They stay current by reading business and environmental magazines and books, attending conferences and seminars, and networking through professional associations. This type of work requires self-motivation, dedication, energy, and decisiveness. Managers and owners often work long hours, sometimes spanning into weeknights and weekends. Commitment to the job and the ability to prioritize are essential.

EXPLORING

Read magazines such as *BusinessWeek* and *Fortune* to keep up with general business news and trends. You can also learn a great deal about green business leaders by subscribing to *Fortune*'s "The Business of Green" video series (http://rss.cnn.com/rss/bizgreen). If there's a particular green business you've been following and are interested in, visit its Web site and contact the company directly to see if you can set up an informational interview (by phone or email) with the owner or manager. Create a list of questions first so you can be prepared to ask them in the interview.

EMPLOYERS

Approximately 402,000 chief executives and 1.72 million general and operations managers were employed in the United States in 2006. Top executives and business managers are employed in most industries, with most finding employment at banks, wholesalers, government agencies, retail establishments, or business service firms. Retail managers usually work within stores and often have irregular hours, especially during special promotions and holidays.

STARTING OUT

An internship or part-time job with a green business is a good way to gain exposure to this field. You can find listings by regularly visiting such employment Web sites as Sustainable Business (http://www.sustainablebusiness.com), Green Job Baron (http://greenjobbaron.com), and Monster (http://www.monster.com).

ADVANCEMENT

Green business managers can advance to higher-level management positions, overseeing more staff and managing more complicated projects. Those with years of experience can start their own companies. Green business owners can expand their companies by branching out into other product and service areas, and by opening branches in other locations. Managers and owners can also share their expertise by consulting with other organizations and associations, speaking at conferences, teaching at universities, as well as writing for various media.

EARNINGS

Salaries for managers and owners of green businesses vary widely, depending on the size and type of company, the location of the business, and the level of job responsibility and number of years on the job. Those who own and manage large, well-known corporations will normally earn significantly higher salaries than those who work for small firms. In 2008 annual salaries for general and operations managers ranged from $45,410 to $91,570 to $166,400 or higher, according to the U.S. Department of Labor. Chief executives averaged between $68,680 and $166,400 or more per year. Those who managed companies and enterprises earned about $128,350 per year. Managers of office and administrative support workers had lower annual salaries that ranged from $27,770 for the lowest 10 percent, to an average of $45,790 for the middle 50 percent, to $74,640 or more for the top 10 percent. The states paying the highest salaries for general and operations managers included Connecticut, Delaware, New Jersey, New York, and Washington.

WORK ENVIRONMENT

Business owners and managers usually work in offices, putting in at least 40 hours per week. They work longer days, sometimes evenings and weekends as well, when they are in the midst of projects and product launches. They are also on call 24/7 so they can be among the first people on hand to help address a potential crisis. Some travel may be involved to meet with business partners, suppliers, and contractors.

OUTLOOK

The U.S. Department of Labor forecasts employment of general and operations managers as showing little or no change through 2016; however, employment in professional, scientific, and technical services is expected to grow faster than the average. And with the growth of interest in green products and sustainable business, green business managers and owners should have fairly decent prospects for employment in the coming years. Competition for top executive positions will be keen, though, so those with strong work backgrounds, leadership skills, and a proven track record in successfully managing companies will have better opportunities to secure work. Also, because of the globalization of industries, individuals with knowledge of international economics, marketing, information systems, and who have the ability to communicate in several languages will have the advantage in the job market.

FOR MORE INFORMATION

For information about certification and to learn more about green business educational programs, contact

Algonquin College
315 Pembroke Street East
Pembroke, Ontario
K8A 3K2
Tel: 613-735-4700
http://www.algonquincollege.com

For information about membership, seminars, publications, and certification, contact

American Management Association
1601 Broadway
New York, NY 10019-7434
Tel: 212-586-8100
Email: customerservice@amanet.org
http://www.amanet.org

For information about professional management certification, contact

Institute of Certified Professional Managers
James Madison University, MSC 5504
Harrisonburg, VA 22807
Tel: 800-568-4120
http://www.icpm.biz

For information about educational forums, conferences, and networking opportunities, contact

National Management Association
2210 Arbor Boulevard
Dayton, OH 45439
Tel: 937-294-0421
Email: nma@nma1.org
http://nma1.org

For information about certification and to learn more about green business educational programs, contact

Sacramento State
College of Continuing Education
3000 State University Drive East
Sacramento, CA 95819-6103
Tel: 916-278-4433
Email: cceinfo@csus.edu
http://www.cce.csus.edu/

Salespeople, Green Product

OVERVIEW

Green product salespeople sell products that help the environment. There are a myriad of things they may sell to consumers, including hybrid cars; organic cleaning supplies (free of chemicals and toxins); organic food, clothing, accessories, and beauty supplies; and energy-saving appliances, such as heating and cooling systems and light fixtures. They sell to customers in person, over the telephone, through various media (print, Web, radio, television), and at conferences and conventions.

HISTORY

Selling products and services is an old profession. Early selling was in the form of trading, and Phoenicians are considered by many to have been among the most successful early traders in history. From about 1550 BCE until the decline of their civilization (around 65 BCE), they developed great wealth through trading primarily with the Greeks. Settled in the coastal region of what is now Lebanon, Phoenicians traded wood, glass, slaves, and a purple powder that was used in dye for pottery. (In ancient Greek, the powder was called *phoinikèia*, from which the Phoenicians derived their name.) They amassed further wealth by traveling widely to obtain other products, such as silver and tin, and by setting up trading posts throughout the Mediterranean.

Starting in the 2nd century B.C. until approximately the late 1500s, caravans of traders traveled through mountains and deserts along the Silk Road, which ran from ancient China to Europe, connecting many civilizations along its path. In the late 13th century, Venetian merchant Marco Polo, who traveled the road for many years and wrote about Asian culture in his book *Million*, gave the road its name because silk was the main commodity traded. Silk was a convenient product to trade because it was lightweight, easy to pack, and in high demand. Other goods traded along the Silk Road included wool, carpets, rugs, tapestries, ivory, spices, ceramics, bronze weapons and other bronze items, mirrors, fragrances, jewels, rice, cotton, and fruits and vegetables. The road became known as Great Silk Road in 1877, when German researcher Ferdinand Richthofen dubbed it as such in his book *China*.

The industrial revolution (beginning in the mid-1700s) created the need for a different mode of selling. Because manufacturing facilities were often in remote locations, representatives were needed to show product samples to merchants and consumers; they became the go-betweens or middlemen. The 1800s saw the rise of the "snake oil salesman," which has evolved into a derogatory term still in use today. Snake oil salesmen traveled the country, using great marketing hype to sell a variety of elixirs and potions that could (supposedly) cure all ailments. Some of these cures may have included snake oil among the ingredients, which is how the term came about. Smart buyers could easily see through the hoax, but there were still plenty of gullible and ailing people who wanted to believe the artfully woven sales pitch and would spend the money anyway. Also in the 1800s, the United States saw the first universal consumer product durable on the market: the cast-iron stove.

According to the 2003 book *Sales Management: A Global Perspective*, salespeople's jobs became more specialized in the 1900s, thanks to new inventions such as the telephone and airplane. Other inventions like the adding machine, typewriter, refrigerator, and vacuum cleaner required a trained sales force to sell the products to consumers. Companies started providing formal training for salesmen, including instructions on how to give "canned" sales presentations (meaning presentations that included the same basic information about certain topics), and assigned specific territories for them to cover. After World War II the sales profession was further elevated due to companies focusing more on customer service—listening to customers and paying attention to their needs—and less on aggressive tactics. Professional, ethical conduct was becoming the norm. Today, sales professionals are recognized and rewarded for their contributions to their companies.

THE JOB

Green product salespeople sell products to people who are interested in conserving energy and resources, and who want to help improve the environment, their community, and their health and well-being. Green products come in all varieties. They may be solar panels for buildings and homes, or compact fluorescent light bulbs for offices. They may be home products such as rain barrels to capture rainwater for gardening, plumbing, or for drinking water (after purification and filtering). They may be clocks made from recycled materials, or body lotions and makeup composed of organic materials (and not tested on animals). To get an idea of the myriad of green products available on the market, visit http://www.greenhome.com, an online store featuring environmentally friendly items.

Green salespeople work on-site in stores; behind the scenes in offices, using computers and telephones to conduct business; or, depending on the job, at a number of locations, traveling to sell products to current and prospective customers at their homes and places of business. Salespeople who represent manufacturers or wholesale distributors sell either one product or a whole line of products. Their clients can include other manufacturers, wholesale and retail establishments, government agencies, and construction contractors.

Years of experience can help a salesperson hone his or her skills, but even new salespeople need to have a thorough understanding and knowledge of the product and company in order to demonstrate the product and communicate its assets to potential customers. They attend internal training programs and company meetings to learn about product development, sales statistics and projections, and to learn new tactics for marketing and promotion. They also stay up to date on new products and their competition by reading trade journals and attending trade shows, conferences, and conventions.

Salespeople who sell technical or scientific products may be teamed up with a technical expert. The salesperson makes the initial contact with the customer, and the technical expert addresses the customers' questions and concerns about the product. The salesperson then closes the deal and keeps in touch with the customer after the purchase, sometimes helping to set up the product and training the customer in how to use it.

Inside sales representatives, as the title suggests, work inside—they stay in one location, spending most of the time on the phone and online, taking orders and helping customers resolve problems with merchandise orders. *Outside sales representatives* travel frequently to meet with current and potential clients. They are usually assigned territories to cover, which can be towns within their state, or multiple states. They spend some time initially talking to customers by phone and emailing with them, discussing the product, its benefits, price, and availability. They set up meetings and bring product samples and catalogs to share in personal meetings as well as in group sales presentations. In addition to knowing their own product and company well, they need to have thorough knowledge of their prospective customers' company and products so they can tailor their sales pitch to address the customers' specific needs.

Pitching is not the sole skill required of successful salespeople: keen listening ability and patience are also essential. Selling products starts with a dialogue between the seller and the customer. Typically, the customer has a problem that needs to be solved; it's up to the salesperson to identify the problem and thoughtfully explain the solution. The ability to put yourself in another person's shoes is extremely helpful in this job. Asking the customer questions, and allowing the customer to speak, is not as easy as it appears. Some customers are not clear communicators. Some may not understand the product and industry at all. Some may even speak a different language—this is where knowledge of other languages and cultures can be beneficial. Salespeople adjust their language and vocabulary to match the customer's mode of communication. In a sense, they treat the dialogue as if they are befriending someone, while still maintaining professionalism and diplomacy throughout.

To keep salespeople motivated and on track, most companies establish sales goals, which are broken down into daily, weekly, monthly, quarterly, and annual quotas that must be met. Goals usually involve making a certain number of phone calls, setting up meetings, and closing a specific number of deals. Salespeople who meet or exceed these goals are recognized with incentives such as commissions on the sales and bonuses.

REQUIREMENTS

High School

For some sales positions, a high school diploma may suffice. A well-rounded liberal arts education provides a solid basis for this job. Take classes in business, math, English, and computers (knowledge of Word, Outlook, Excel, and PowerPoint is beneficial).

Postsecondary Training

Companies may require a bachelor's degree for some sales and sales management positions. Course work in business, marketing, finance, and English is useful in this field. Salespeople interact heavily with others: A large part of the job is listening to what people need and tailoring sales pitches to address those needs. Classes in psychology, sociology, writing, and speech will help hone communication skills.

Certification or Licensing

Certification is not required, although it can help advance a salesperson's career. The National Association of Sales Professionals offers certification for three levels of demonstrated sales skills: level one is for the demonstration of sales skills; level two is for influence skills; and level three is for strategic skills.

Other Requirements

Successful salespeople enjoy working with people; otherwise, they would not (and should not) be doing what they do. Customer service is a large part of the job, and being a good listener and clear communicator helps tremendously. Patience and diplomacy are also key elements in dealing with people, and especially in dealing with difficult customers. In some positions, the ability to speak another language and knowledge of other cultures is beneficial. Some salespeople cover large territories, traveling frequently to different states to meet with prospective clients. They are often away from home for several days at a time. Having a flexible attitude and understanding family members can help ease the stress of this type of work schedule.

EXPLORING

Talk to salespeople to learn more about the job. The next time you're in a department or specialty store, if it's not too busy, ask the salesperson how he or she got the job, and what they like most and least about it. You can also find out more about the types of sales jobs available by visiting employment Web sites such as Monster (http://www.monster.com), Sales Jobs (http://www.salesjobs.com), TreeHugger (http://jobs.treehugger.com), and Sustain Lane (http://www.sustainlane.com/green-jobs).

EMPLOYERS

According to the U.S. Department of Labor, approximately 4.5 million salespeople held jobs in retail in 2006. Many work for department stores and specialty shops, and some are self-employed, representing direct-sales companies and mail-order houses. Green product salespeople may work for companies that produce solar panels, for example, or pollution-control devices, or building materials made from recycled products. They may represent wholesalers and manufacturers, selling products to individual businesses. They may even specialize in selling advertising space for green companies. The largest employers of retail salespersons are department stores, clothing and accessories stores, building material and garden equipment and supplies dealers, other general merchandise stores, and motor vehicle and parts dealers.

STARTING OUT

A part-time or summer job in a department store, or green product store if one is near you, is the ideal way to see if this type of work is a good fit for you. Check your local newspaper's classified listings for employment openings. If you have a green store in mind that you would like to work for, visit its Web site and look for the "employment" or "work for us" section to see if there are openings for sales associates. You can also visit the store in person and ask to fill out an employment application.

ADVANCEMENT

Salespeople with proven track records and years of experience in the field can advance to senior salesperson and management positions. They can move up to manage larger sales teams and departments. Those who are on staff at companies can advance by starting their own companies. Those who are self-employed can grow their companies by hiring more staff, picking up more clients, providing other services and products, and opening other companies. Salespeople with excellent sales records who are well known in the field can lecture, give motivational talks, teach at universities, and write for various print and online publications.

EARNINGS

In 2008 retail salespeople earned annual salaries that ranged from $15,340 to $39,810. Those who worked for clothing and department stores had the lowest salaries, averaging around $21,150, while automobile salespeople topped out at $42,410 per year. Sales reps for wholesalers and manufacturers earned higher salaries, with the lowest 10 percent averaging $26,950, the middle 50 percent earning about $51,330, and the top 10 percent taking home $106,040 or more per year.

Salary structures vary depending on how the company sets them up. Many salespeople earn base salaries and commissions for the sales they make. Others earn an hourly wage and then a percentage of the sales. To keep salespeople motivated and on top of sales goals, companies may offer other incentives as well, including bonuses, prizes and awards, banquets, and profit-sharing plans, in addition to discounts on company services and products.

WORK ENVIRONMENT

Salespeople work in stores, either on the sales floor or behind the scenes in offices. Some work outdoors if they sell automobiles or products such as garden supplies and equipment. They work either part-time hours or, if they are on staff, 40 or more hours per week. Shifts will vary and salespeople are usually scheduled to work longer hours during sales, holidays, and when the store is closed for inventory review.

OUTLOOK

Employment opportunities for retail, manufacturer, and wholesale salespeople are expected to grow about as fast as the average for all occupations through 2016, according to the U.S. Department of Labor.

Salespeople in general should have decent opportunities to find work in the next few years, providing the economy continues to stabilize. The sales industry is heavily dependent on the economy. In times of economic recession and downturns, people are more cautious with their money and spend less, and thus sales decline. In

2008, however, during the global economic slowdown, the Natural Marketing Institute reported that, based on analysis of green consumers and their shopping habits, green and healthy products accounted for $209 billion in sales, and predicted that figure to reach more than $400 billion by 2010. Green products offer consumers opportunities to conserve energy and save money in the long run, and so green product salespeople may have better opportunities to find work, regardless of what's happening with the economy.

FOR MORE INFORMATION

For publications and upcoming events about marketing and advertising, contact

American Marketing Association
311 South Wacker Drive, Suite 5800
Chicago, IL 60606-6629
Tel: 312-542-9000
http://www.marketingpower.com

For information about certification, and to search job listings, visit

National Association of Sales Professionals
37557 Newburgh Park Circle
Livonia, MI 48152-1373
Tel: 248-230-4303
http://www.nasp.com

For information about conferences, membership, and professional education in retailing, contact

National Retail Federation
325 7th Street, NW, Suite 1100
Washington, DC 20004-2825
Tel: 202-783-7971
http://www.nrf.com

For information about green business research studies, visit NMI's Web site.

Natural Marketing Institute (NMI)
11200 Rockville Pike, Suite 504
Rockville, MD 20852-3189
Tel: 240-747-3000
http://www.marketresearch.com

Surveyors

OVERVIEW

Surveyors mark exact measurements and locations of elevations, points, lines, and contours on or near Earth's surface. They measure distances between points to determine property boundaries and to provide data for mapmaking, construction projects, and other engineering purposes. There are approximately 148,000 surveyors, *cartographers, photogrammetrists,* and *surveying technicians* employed in the United States. Of those, about 60,000 are surveyors and about 12,000 are cartographers and photogrammetrists.

HISTORY

As the United States expanded from the Atlantic to the Pacific, people moved over the mountains and plains into the uncharted regions of the West. They found it necessary to chart their routes and to mark property lines and borderlines by surveying and filing claims.

The need for accurate geographical measurements and precise records of those measurements has increased over the years. Surveying measurements are needed to determine the location of a trail, highway, or road; the site of a log cabin, frame house, or skyscraper;

the right-of-way for water pipes, drainage ditches, and telephone lines; and for the charting of unexplored regions, bodies of water, land, and underground mines.

As a result, the demand for professional surveyors has grown and become more complex. New computerized systems are now used to map, store, and retrieve geographical data more accurately and efficiently. This new technology has not only improved the process of surveying but extended its reach as well. Surveyors can now make detailed maps of ocean floors and the moon's surface.

THE JOB

On proposed construction projects, such as highways, airstrips, and housing developments, it is the surveyor's responsibility to make necessary measurements through an accurate and detailed survey of the area. The surveyor usually works with a field party consisting of several people. Instrument assistants, called *surveying and mapping technicians*, handle a variety of surveying instruments including the theodolite, transit, level, surveyor's chain, rod, and other electronic equipment. In the course of the survey, it is important that all readings be recorded accurately and that field notes be maintained so that the survey can be checked for accuracy.

Surveyors may specialize in one or more particular types of surveying.

Land surveyors establish township, property, and other tract-of-land boundary lines. Using maps, notes, or actual land title deeds, they survey the land, checking for the accuracy of existing records. This information is used to prepare legal documents such as deeds and leases. *Land surveying managers* coordinate the work of surveyors, their parties, and legal, engineering, architectural, and other staff involved in a project. In addition, these managers develop policy, prepare budgets, certify work upon completion, and handle numerous other administrative duties.

Highway surveyors establish grades, lines, and other points of reference for highway construction projects. This survey information is essential to the work of the numerous engineers and the construction crews who build the new highway.

Geodetic surveyors measure large masses of land, sea, and space that must take into account the curvature of Earth and its geophysical characteristics. Their work is helpful in establishing points of reference for smaller land surveys, determining national boundaries, and preparing maps. Geodetic computers calculate latitude, longitude, angles, areas, and other information needed for mapmaking.

They work from field notes made by an engineering survey party and also use reference tables and a calculating machine or computer.

Marine surveyors measure harbors, rivers, and other bodies of water. They determine the depth of the water through measuring sound waves in relation to nearby land masses. Their work is essential for planning and constructing navigation projects such as breakwaters, dams, piers, marinas, and bridges, and for preparing nautical charts and maps.

Mine surveyors make surface and underground surveys, preparing maps of mines and mining operations. Such maps are helpful in examining underground passages within the levels of a mine and assessing the volume and location of raw material available.

Geophysical prospecting surveyors locate and mark sites considered likely to contain petroleum deposits. *Oil-well directional surveyors* use sonic, electronic, and nuclear measuring instruments to gauge the presence and amount of oil- and gas-bearing reservoirs. *Pipeline surveyors* determine rights-of-way for oil construction projects, providing information essential to the preparation for and laying of the lines.

Photogrammetric engineers determine the contour of an area to show elevations and depressions and indicate such features as mountains, lakes, rivers, forests, roads, farms, buildings, and other landmarks. Aerial, land, and water photographs are taken with special equipment able to capture images of very large areas. From these pictures, accurate measurements of the terrain and surface features can be made. These surveys are helpful in construction projects and in the preparation of topographical maps. Photogrammetry is particularly helpful in charting areas that are inaccessible or difficult to travel.

Surveyors use the Global Positioning System (GPS) to locate reference points. GPS is a worldwide radio-navigation system formed by a constellation of 24 satellites and ground stations. The GPS technology is so precise that it enables surveyors to measure reference points within a centimeter.

Surveyors, cartographers and photogrammetrists, and surveying and mapping technicians also use Geographic Information Systems (GIS) in their work. GIS is a computer system used to capture, assemble, integrate, analyze, and display data about geographical locations in a digital format. Workers typically use GIS to make maps for environmental studies, geology, engineering, planning, business marketing, and other disciplines. In fact, *geographic information specialist* has become the recognized job title for a mapping specialist.

Kriste Lindberg, a cave expert and director of Indiana Karst Conservancy, Inc., who was one of the principal surveyors of the Lost River cave system, works her way through an area of the Lost River Cave. The cave is filled with rare and endangered species, such as the blind cave fish. *AP Photo/The Indianapolis Star, Charlie Nye*

REQUIREMENTS
High School
You can prepare for this field by taking plenty of math and science courses in high school. Take algebra, geometry, and trigonometry to become comfortable making different calculations. Earth science, chemistry, and physics classes should also be helpful. Geography will help you learn about different locations, their characteristics, and cartography. Benefits from taking mechanical drawing and other drafting classes include an increased ability to visualize abstractions, exposure to detailed work, and an understanding of perspectives. Taking computer science classes will prepare you for working with technical surveying equipment.

Postsecondary Training
Most surveyors hold a bachelor's degree in surveying. Cartographers and photogrammetrists usually have a bachelor's degree in cartography, geography, surveying, engineering, forestry, computer science, or a physical science. Many universities offer bachelor's degree

programs in surveying. Those interested in surveying technology will also find programs offered at community colleges, technical institutes, and vocational schools.

Certification or Licensing

All 50 states require that surveyors making property and boundary surveys be licensed or registered. The requirements for licensure vary, but most require a degree in surveying or a related field, a certain number of years of experience, and passing examinations given by the National Council of Examiners for Engineering and Surveying (NCEES). After passing the first exam, candidates usually work for four years under the supervision of an experienced surveyor and then take a second exam to become fully licensed. In addition to licensure requirements, many states also have continuing education requirements for surveying professionals. Also, many states now require that cartographers and photogrammetrists be licensed as surveyors; some states now have specific licenses for photogrammetrists. If you are seeking employment in the federal government, you must take a civil service examination and meet the educational, experience, and other specified requirements for the position.

To advance their careers, surveying technicians can get voluntary certification from the National Society of Professional Surveyors. The American Society for Photogrammetry and Remote Sensing also offers voluntary certification in photogrammetry, remote sensing, and GIS.

Other Requirements

The ability to work with numbers and perform mathematical computations accurately and quickly is very important. Other helpful qualities are the ability to visualize and understand objects in two and three dimensions (spatial relationships) and the ability to discriminate between and compare shapes, sizes, lines, shadings, and other forms (form perception).

Surveyors walk a great deal and carry equipment over all types of terrain, so endurance and coordination are important physical assets. In addition, surveyors direct and supervise the work of their team, so you should have clear communication skills, be good at working with other people, and demonstrate leadership abilities.

EXPLORING

While you are in high school, begin to familiarize yourself with terms, projects, and tools used in this profession by reading books and magazines on the topic. One magazine that is available online is *Professional Surveyor Magazine* (http://www.profsurv.com). One of the best opportunities for experience is a summer job with a construction outfit or company that requires survey work. Even if the job does not involve direct contact with survey crews, you will have the opportunity to observe surveyors and talk with them about their work.

Some colleges have work-study programs that offer on-the-job experience. These opportunities, like summer or part-time jobs, provide helpful contacts in the field that may lead to future full-time employment. If your college does not offer a work-study program and you can't find a paying summer job, consider volunteering at an appropriate government agency. The U.S. Geological Survey and the Bureau of Land Management usually have volunteer opportunities in select areas.

You can learn more about GPS by taking a GPS tutorial at this Web site: http://www.trimble.com/gps/index.shtml. Read up on GIS by visiting http://www-sul.stanford.edu/depts/gis/whatgis.html#other.

EMPLOYERS

According to the U.S. Department of Labor, more than two-thirds of surveying workers in the United States are employed in engineering, architectural, and surveying firms. Federal, state, and local government agencies are the next largest employers of surveying workers. Federal government employers are the U.S. Geological Survey, the Bureau of Land Management, the National Geodetic Survey, the National Geospatial Intelligence Agency, and the U.S. Army Corps of Engineers. Surveying professionals who work for state and local governments work for highway departments or urban planning and redevelopment agencies. The majority of the remaining surveyors work for construction firms, oil and gas extraction companies, and mining and utility companies. Only a small number of surveyors are self-employed.

STARTING OUT

Apprentices with a high school education can enter the field as equipment operators or surveying assistants. Those who have post-secondary education can enter the field more easily by beginning as surveying and mapping technicians.

College graduates can learn about job openings through their schools' placement services or through potential employers that may visit their campus. Many cities have employment agencies that specialize in seeking out workers for positions in surveying and related fields. Check your local newspaper or telephone book to see if such recruiting firms exist in your area.

ADVANCEMENT

With experience, workers advance through the leadership ranks within a surveying team. Workers begin as assistants and then can move into positions such as senior technician, party chief, and, finally, licensed surveyor. Because surveying work is closely related to other fields, surveyors can move into electrical, mechanical, or chemical engineering or specialize in drafting.

EARNINGS

According to the U.S. Department of Labor, surveyors earned a median annual salary of $52,980 in 2008. The lowest 10 percent earned $29,600 and the highest 10 percent earned $85,620 a year. In general, the federal government paid the highest wages to its surveyors, $78,710 a year in 2008.

Surveying and mapping technicians earned salaries in that same year that ranged from $21,690 to $58,030; and cartographers and photogrammetrists' annual salaries ranged from $31,440 to $78,710.

Most positions with the federal, state, and local governments and with private firms provide life and medical insurance, pension, vacation, and holiday benefits.

WORK ENVIRONMENT

Surveyors work 40-hour weeks except when overtime is necessary to meet a project deadline. The peak work period is during the summer months when weather conditions are most favorable. It is not uncommon, however, for surveyors to be exposed to adverse weather conditions.

Some survey projects may involve hazardous conditions, depending on the region and climate as well as the plant and animal life. Survey crews may encounter snakes, poison ivy, and other hazardous plant and animal life, and they may suffer heat exhaustion, sunburn, and frostbite while in the field. Survey projects, particularly those near construction projects or busy highways, may impose dangers of injury from heavy traffic, flying objects, and other accidental hazards. Unless the surveyor is employed only for office assignments, the work location most likely will change from survey to survey. Some assignments may require the surveyor to be away from home for periods of time.

OUTLOOK

The U.S. Department of Labor predicts the employment of surveyors to grow much faster than the average occupation through 2016. Increased demand for geographic data for natural resource exploration, urban planning, construction, and other applications will mean more jobs for surveyors. The outlook is best for those who have college degrees and advanced field experience. While Global Positioning System and Geographic Information Systems have decreased the size of surveying teams, those who have strong technical and computer skills will have the edge in the job market.

Growth in urban and suburban areas (with the need for new streets, homes, shopping centers, schools, gas and water lines) will provide employment opportunities. State and federal highway improvement programs and local urban redevelopment programs will also provide jobs for surveyors. The expansion of industrial and business firms and the relocation of some firms to large undeveloped tracts will also create job openings. Construction projects are closely tied to the state of the economy, so employment may fluctuate from year to year.

FOR MORE INFORMATION

For information on awards and recommended books to read, contact the following organizations or check out their Web sites:

American Association for Geodetic Surveying
6 Montgomery Village Avenue, Suite 403
Gaithersburg, MD 20879-3546
Tel: 240-632-9716
Email: aagsmo@acsm.net
http://www.aagsmo.org

National Society of Professional Surveyors
6 Montgomery Village Avenue, Suite 403
Gaithersburg, MD 20879-3557
Tel: 240-632-9716
Email: trisha.milburn@acsm.net
http://www.nspsmo.org

For information on state affiliates and colleges and universities offering land surveying programs, contact

American Congress on Surveying and Mapping
6 Montgomery Village Avenue, Suite 403
Gaithersburg, MD 20879-3546
Tel: 240-632-9716
Email: info@acsm.net
http://www.acsm.net

For information on photogrammetry and careers in the field, contact

American Society for Photogrammetry and Remote Sensing
5410 Grosvenor Lane, Suite 210
Bethesda, MD 20814-2160
Tel: 301-493-0290
Email: asprs@asprs.org
http://www.asprs.org

For information on volunteer opportunities with the federal government, contact the following agencies:

Bureau of Land Management
Office of Public Affairs
1849 C Street, NW, Room 5665
Washington, DC 20240
Tel: 202-208-3801
http://www.blm.gov

U.S. Geological Survey
12201 Sunrise Valley Drive
Reston, VA 20192-0002
Tel: 703-648-4000
http://www.usgs.gov

Further Reading

Asafu-Adjaye, John. *Environmental Economics for Non-Economists: Techniques and Policies for Sustainable Development*. Hackensack, N.J.: World Scientific Publishing Company, 2005.

Association of Manufacturing Excellence. *Green Manufacturing: Case Studies in Leadership and Improvement*. New York: Productivity Press, 2007.

Batie, Clarence M. *The Economists: Profiles and Careers*. Boston: Pearson Custom Publishing, 2006.

Brewer, Richard. *Conservancy: The Land Trust Movement in America*. Hanover, N.H.: Dartmouth, 2003.

Christian, Jeffrey E. *The Headhunter's Edge*. New York: Random House, 2002.

Dallas, Nick. *Green Business Basics: 24 Lessons for Meeting the Challenges of Global Warming*. New York: McGraw-Hill, 2008.

Dines, Nicholas and Kyle Brown. *Landscape Architect's Portable Handbook*. New York: McGraw-Hill Professional, 2001.

Entrepreneur Press. *Start Your Own Executive Recruiting Business*. Newburgh, N.Y.: Entrepreneur Press, 2007.

Esty, Daniel and Andrew Winston. *Green to Gold: How Smart Companies Use Environmental Strategy to Innovate, Create Value, and Build Competitive Advantage*. Revised edition. Hoboken, N.J.: Wiley, 2009.

Finlay, William and James E. Coverdill. *Headhunters: Matchmaking in the Labor Market*. Ithaca, N.Y.: ILR Press, 2007.

Friend, Gil. *The Truth about Green Business*. Saddle River, N.J.: FT Press, 2009.

Johnston, David, and Scott Gibson. *Green from the Ground Up: Sustainable, Healthy, and Energy-Efficient Home Construction*. Newtown, Conn.: Taunton, 2008.

Jones, Van. *The Green Collar Economy*. New York: HarperOne, 2009.

Keiter, Robert B. *Keeping Faith with Nature: Ecosystems, Democracy, and America's Public Lands*. New Haven, Conn.: Yale University Press, 2003.

Kiernan, Matthew J. *Investing in a Sustainable World: Why GREEN Is the New Color of Money on Wall Street*. New York: AMACOM Books, 2008.

Phyper, John. *Good to Green: Managing Business Risks and Opportunities in the Age of Environmental Awareness*. Hoboken, N.J.: John Wiley & Sons, 2009.

Sullivan, Rory. *Corporate Responses to Climate Change: Achieving Emissions Reductions through Regulation, Self-Regulation and Economic Incentives.* Sheffield, U.K.: Greenleaf Publishing, 2008.

Swain, Ann and Jane Newell-Brown. *The Professional Recruiter's Handbook: Delivering Excellence in Recruitment Practice.* London: Kogan Page, 2009.

Swonk, Diane. *The Passionate Economist: Finding the Power and Humanity Behind the Numbers.* Hoboken, N.J.: Wiley, 2003.

Van Lengen, Johan. *The Barefoot Architect: A Handbook for Green Building.* Bolinas, Calif.: Shelter Publications, 2007.

Vernon, Siobhan. *Landscape Architect's Pocket Book.* Burlington, Mass.: Elsevier Science & Technology Books, 2009.

Werthmann, Christian. *Green Roofs Gardens: A Case Study.* New York: Princeton Architectural Press, 2007.

Woodson, R. *Be A Successful Green Builder.* New York: McGraw-Hill Professional, 2008.

Yudelson, Jerry. *Green Building A to Z: Understanding the Language of Green Building.* Gabriola Island, Canada: New Society Publishers, 2007.

Index